Wireless Medical Systems
and Algorithms

DESIGN AND APPLICATIONS

Devices, Circuits, and Systems

Series Editor
Krzysztof Iniewski
Emerging Technologies CMOS Inc.
Vancouver, British Columbia, Canada

PUBLISHED TITLES:

PUBLISHED TITLES:

Reconfigurable Logic: Architecture, Tools, and Applications
Pierre-Emmanuel Gaillardon

Semiconductor Radiation Detection Systems
Krzysztof Iniewski

Smart Grids: Clouds, Communications, Open Source, and Automation
David Bakken

Smart Sensors for Industrial Applications
Krzysztof Iniewski

Soft Errors: From Particles to Circuits
Jean-Luc Autran and Daniela Munteanu

Solid-State Radiation Detectors: Technology and Applications
Salah Awadalla

Technologies for Smart Sensors and Sensor Fusion
Kevin Yallup and Krzysztof Iniewski

Telecommunication Networks
Eugenio Iannone

Testing for Small-Delay Defects in Nanoscale CMOS Integrated Circuits
Sandeep K. Goel and Krishnendu Chakrabarty

VLSI: Circuits for Emerging Applications
Tomasz Wojcicki

Wireless Medical Systems and Algorithms: Design and Applications
Pietro Salvo and Miguel Hernandez-Silveira

Wireless Technologies: Circuits, Systems, and Devices
Krzysztof Iniewski

Wireless Transceiver Circuits: System Perspectives and Design Aspects
Woogeun Rhee

FORTHCOMING TITLES:

Advances in Imaging and Sensing
Shuo Tang and Daryoosh Saeedkia

Circuits and Systems for Security and Privacy
Farhana Sheikh and Leonel Sousa

Magnetic Sensors: Technologies and Applications
Laurent A. Francis and Kirill Poletkin

MRI: Physics, Image Reconstruction, and Analysis
Angshul Majumdar and Rabab Ward

Multisensor Attitude Estimation: Fundamental Concepts and Applications
Hassen Fourati and Djamel Eddine Chouaib Belkhiat

FORTHCOMING TITLES:

Wireless Medical Systems
and Algorithms

DESIGN AND APPLICATIONS

Edited by
Pietro Salvo
University of Pisa, Italy

Miguel Hernandez-Silveira
Sensium Healthcare LTD, Abingdon, UK

Krzysztof Iniewski MANAGING EDITOR
Emerging Technologies CMOS Inc.
Vancouver, British Columbia, Canada

CRC Press
Taylor & Francis Group
Boca Raton London New York

CRC Press is an imprint of the
Taylor & Francis Group, an **informa** business

CRC Press
Taylor & Francis Group
6000 Broken Sound Parkway NW, Suite 300
Boca Raton, FL 33487-2742

First issued in paperback 2018

© 2016 by Taylor & Francis Group, LLC
CRC Press is an imprint of Taylor & Francis Group, an Informa business

No claim to original U.S. Government works

ISBN-13: 978-1-4987-0076-4 (hbk)
ISBN-13: 978-1-138-58500-3 (pbk)

Visit the Taylor & Francis Web site at
http://www.taylorandfrancis.com

and the CRC Press Web site at
http://www.crcpress.com

Contents

SECTION I Technologies and Manufacturing

SECTION II Algorithms and Data Processing

Preface

The most recent research efforts in medical therapies and monitoring aim to develop wearable and wireless devices that can help assess patients' health conditions during everyday life. The advantages of these devices are twofold: avoiding patients' hospitalization to reduce the costs for the healthcare systems and improve patients' comfort and providing long-term continuous monitoring of patients' physiological parameters for achieving personalized therapies and preventing potentially life-threatening events. During the last decade, the challenge has been to merge disciplines such as chemistry, biology, engineering, and medicine to produce a new class of smart devices that are capable of managing and monitoring a wide range of cognitive and physical disabilities. Within this research frame, the medical market has started to offer sophisticated medical devices combined with wireless communication capabilities. These systems provide caregivers with new opportunities to access patients' medical data in real time and enhance the possibilities of prompt interventions in case of emergency.

This book tries to cover the most important steps that lead to the development and fabrication of wireless medical devices and some of the most advanced algorithms and data processing in the field. The book chapters will provide readers with an overview and practical examples of some of the latest challenges and solutions for medical needs, electronic design, advanced materials chemistry, wireless body sensors networks, and technologies suitable for wireless medical devices. The book includes practical examples and case studies for readers with different skills, ranging from engineers to medical doctors, chemists, and biologists, who can find new exciting ideas and opportunities for their work.

In the first section, we describe the technological and manufacturing challenges for the development of wireless medical devices. The first two chapters discuss the development and fabrication of electronics and packaging of biochips with emphasis on the readout circuit and microassembly. The other two chapters report on research studies and devices for wireless biomedical sensing.

In the second section, readers are introduced to the techniques and strategies that can optimize the performances of algorithms for medical applications and provide robust results in terms of data reliability. Two chapters are dedicated to practical examples in the field of brain–computer interfaces and artificial pancreas.

We thank all the authors for their contributions to this volume. Finally, we thank the staff of CRC Press, Boca Raton, FL, for their kind help in publishing this book.

MATLAB® is a registered trademark of The MathWorks, Inc. For product information, please contact:

The MathWorks, Inc.
3 Apple Hill Drive
Natick, MA 01760-2098 USA
Tel: 508 647 7000
Fax: 508-647-7001
E-mail: info@mathworks.com
Web: www.mathworks.com

Editors

Pietro Salvo, PhD, earned his MSc degree in electronics engineering and his PhD degree in automation, robotics and bioengineering from the University of Pisa, Italy, in 2004 and 2009, respectively. He was a research associate at the Institute for the Conservation and Promotion of Cultural Heritage, National Council of Research, Florence, Italy, from 2008 to 2009. From 2009 to 2013, he was at the Centre for Microsystems Technology (CMST), Ghent University, Belgium, where he was responsible for the development of sensors, and electronic and microfluidic systems for bioengineering applications. Presently, he is with the Department of Chemistry and Industrial Chemistry, University of Pisa, Italy. His current research activity aims at the development of noninvasive tools and sensors for diagnostics and monitoring in healthcare applications.

Miguel Hernandez-Silveira was born in Caracas, Venezuela, in 1970. He earned his computer science and electronics engineering degree from the Universidad Fermin Toro, Venezuela, in 1996. He joined one of the most prestigious Venezuelan universities situated in the Andes region (UNET) in 1998, where he held lecturer and researcher positions until 2003. He was a member, a cofounder, and the director of the Biomedical Engineering Research Group of this institution. During his time in UNET, Miguel participated in the design and development of different medical devices and technologies for healthcare monitoring involving biomedical circuits, systems, and algorithms. He earned a PhD degree in biomedical engineering from the University of Surrey (Guildford, UK), where he worked in the development of electrode technologies for functional electrical stimulation to assist the gait of patients with upper motor neuron impairments.

Miguel joined Sensium Healthcare Ltd (formerly Toumaz) in 2008, where his main role has been in the development and optimization of DSP and machine learning vital-signs algorithms for ultralow-power Sensium® microchips. He is currently the head of Biomedical Technologies of the Toumaz Group, and together with his team, he actively participates in and coordinates the ongoing development of new intelligent algorithms for wireless healthcare monitoring. Dr. Hernandez-Silveira is also a visiting researcher at Imperial College of London, where he occasionally imparts lectures, supervises postgraduate projects, and participates/contributes in large-scale research projects at the Centre for Bio-inspired Technologies.

His main interests include wireless low-power healthcare systems and smart algorithms for analysis and interpretation of physiological data. Miguel has been either the author or a coauthor of publications in this field.

Krzysztof (Kris) Iniewski is managing R&D at Redlen Technologies Inc., a start-up company in Vancouver, Canada. Redlen's revolutionary production process for advanced semiconductor materials enables a new generation of more accurate, all-digital, radiation-based imaging solutions. Kris is also a founder of ET CMOS Inc. (www.etcmos.com), an organization of high-tech events covering communications, microsystems, optoelectronics, and sensors. In his career, Dr. Iniewski held numerous faculty and management positions at the University of Toronto, University of Alberta, Simon Fraser University, and PMC-Sierra Inc. He has published more than 100 research papers in international journals and conferences. He holds 18 international patents granted in the United States, Canada, France, Germany, and Japan. He is a frequent invited speaker and has consulted for multiple organizations internationally. He has written and edited several books for CRC Press, Cambridge University Press, IEEE Press, Wiley, McGraw-Hill, Artech House, and Springer. His personal goal is to contribute to healthy living and sustainability through innovative engineering solutions. He can be reached at kris.iniewski@gmail.com.

Contributors

Su-Shin Ang
Sensium Healthcare
Abingdon, United Kingdom

Ahmet F. Coskun
Electrical Engineering Department
University of California
Los Angeles, California

Yuan Gao
Department of Electrical and Computer
    Engineering
National University of Singapore
and
Institute of Microelectronics
Singapore, Singapore

Pantelis Georgiou
Centre for Bio-Inspired Technology
Imperial College London
London, United Kingdom

Maysam Ghovanloo
Georgia Institute of Technology
Atlanta, Georgia

Chun-Huat Heng
Department of Electrical and Computer
    Engineering
National University of Singapore
and
Institute of Microelectronics
Singapore, Singapore

Miguel Hernandez-Silveira
Sensium Healthcare
Abingdon, United Kingdom

and

Department of Electrical and Electronic
    Engineering
Imperial College of London
South Kensington Campus
London, United Kingdom

Pau Herrero
Centre for Bio-Inspired Technology
Imperial College London
London, United Kingdom

Heikki Karvonen
Centre for Wireless Communications
University of Oulu
Oulu, Finland

Ana Matran-Fernadez
School of Computer Science and
    Electronic Engineering
University of Essex
Essex, United Kingdom

Onur Mudanyali
Electrical Engineering Department
University of California
Los Angeles, California

Nick Oliver
St. Mary's Campus
Imperial College London Medical
    School
London, United Kingdom

Aydogan Ozcan
Electrical Engineering Department
University of California
Los Angeles, California

Hangue Park
Georgia Institute of Technology
Atlanta, Georgia

Peter Pesl
Centre for Bio-Inspired Technology
Imperial College London
London, United Kingdom

Juha Petäjäjärvi
Centre for Wireless Communications
University of Oulu
Oulu, Finland

Riccardo Poli
School of Computer Science and
    Electronic Engineering
University of Essex
Essex, United Kingdom

Monika Reddy
St. Mary's Campus
Imperial College London Medical
    School
London, United Kingdom

Saul Rodriguez
KTH Royal Institute of Technology
Stockholm, Sweden

Ana Rusu
KTH Royal Institute of Technology
Stockholm, Sweden

Dimitrios Soudris
Microprocessors and Digital Systems
    Laboratory
Electrical and Computer Engineering
    Department
National Technical University of Athens
Athens, Greece

Serguei Stoukatch
Microsys Lab
University of Liege
Liege, Belgium

Sha Tao
KTH Royal Institute of Technology
Stockholm, Sweden

Vasileios Tsoutsouras
Microprocessors and Digital Systems
    Laboratory
Electrical and Computer Engineering
    Department
National Technical University of Athens
Athens, Greece

Davide Valeriani
School of Computer Science and
    Electronic Engineering
University of Essex
Essex, United Kingdom

Maria Xenou
St. Mary's Campus
Imperial College London Medical
    School
London, United Kingdom

Sotirios Xydis
Microprocessors and Digital Systems
    Laboratory
Electrical and Computer Engineering
    Department
National Technical University of Athens
Athens, Greece

Hongying Zhu
Electrical Engineering Department
University of California
Los Angeles, California

Section I

Technologies and Manufacturing

1 Advances in Technologies for Implantable Bioelectronics

Saul Rodriguez, Sha Tao, and Ana Rusu

CONTENTS

1.1 INTRODUCTION

The increasing capability in the electronics industry, as underlined by Moore's law, has been the engine for technological development, driving this industry for the last 40 years. This engine has enabled a wide set of technologies for sensing, actuating, processing, and communicating with the environment and opens the door for new applications by progressively creating smaller, smarter, and always-connected devices. The wireless communications and electronics technologies have already revolutionized the way we communicate. Now, wireless communications and electronics face a new need—sustainable society. Toward a long-term response to sustainable society, the researchers started investigating technologies for future healthcare systems. Recent advances in biosensing and material technologies, nano-/microelectronics, and wireless communications have enabled the engineering of implantable biomedical devices for monitoring biosignals. These technologies have the potential to revolutionize medical care by considerably decreasing the healthcare costs and improving the patient's quality of life. Providing prevention-oriented and personalized healthcare services has become a necessity to keep people away from diseases and to reduce the healthcare costs. To implement such kind of healthcare systems, portable, wearable, and implantable monitoring biomedical devices are required. The development of these biomedical devices is a very challenging task because of their stringent requirements for ultra-low power and packaging in a small form factor. The recent research efforts in bioelectronics (the integration of biology with electronics) show encouraging results toward low-cost, ultra-low-power biomedical devices that will lead to ubiquitous wireless healthcare, continuing to improve our quality of life. In terms of hardware implementation, different components (e.g., readout circuit, communication, and digital processing) of a biomedical device dissipate different levels of power depending on the mode of operation (sleep, wake-up, and data acquisition) and the medical application. Therefore, the optimization of biosensor architecture as well as its building blocks and circuits in terms of power is critical. Implantable bioelectronics is one of the most fascinating areas that can benefit from the rapid development of bioelectronics and all related technologies. The implantable bioelectronics technology is greatly advanced for some medical applications, such as pacemakers and prosthetic devices, but extensive research efforts are focused also on many other medical applications, such as insulin pumps, implantable neural recording, gastric stimulators, cochlear implants, retinal implants, and so on. In all these applications, one of the most critical issues is the interface between living tissues and the implantable electronic devices. Therefore, the development of implantable biomedical devices is very challenging since extreme constraints on weight, volume, power, and biocompatibility are added to the existing stringent limitations of noninvasive biosensors. Additionally, the device's lifetime should be as long as possible; in an ideal case, it should exceed the lifetime of the host. Thus, it is highly desirable to have self-powered operation in implantable biosensors, as the lifetime of these devices would be then extended and the overall cost would be reduced. The development of ultra-low-power circuits, highly efficient energy harvesting, storage, and power management technologies is the most promising strategy to overcome the challenges and obstacles presented by the self-powered implantable biomedical devices.

The generic block diagram of a biomedical device is shown in Figure 1.1. The core components of such device include the sensing, readout, communication, signal processing, and power management modules. The sensing module is composed of a sensor/actuator, which interfaces with the biomedical environment. For implants, the

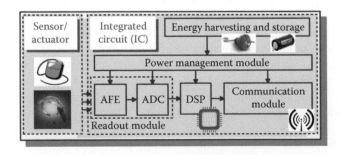

FIGURE 1.1   A generic block diagram of a biomedical device.

environment is inside the human body. The sensing power varies with the application or the complexity of detecting a certain event. The readout module, composed of an analog front-end (AFE) and an analog-to-digital converter (ADC), captures and processes the sensed signal and converts it to a digital domain, where it is further processed by the digital signal processor (DSP). The communication module provides communication between the biomedical device and the outside world, and it is usually a wireless transceiver (Bluetooth, ZigBEE, etc.). Additionally, a power management module is required to efficiently power the different modules of the biomedical device.

This chapter focuses on the implantable readout module, which provides the direct interface with the biological tissues and is strongly dependent on the type of biosignal. Implantable readout circuits for biopotential and electrochemical sensing are introduced, and their challenges are briefly described. State-of-the-art ADC architectures for implantable readouts are then presented and a test-case ADC for neural recording applications is introduced. Additionally, the powering issues and challenges are covered and possible solutions for self-powered implantable biosensors are briefly presented.

1.2 READOUT CIRCUITS

The implantable readout circuits extract information from a variety of biosensors within the human body, providing the interface between biosignals and DSP. The majority of implantable readout circuits can be classified into two big groups: readout circuits for biopotential sensing and readout circuits for electrochemical biosensing. Biopotential sensing consists of recording the electrical activity in tissues that is caused by excitations at the nervous system. Electrochemical biosensing, on the other hand, consists of measuring electrical subproducts of an intentional electrochemical reaction, which involves physiological substances within the body and functionalized materials on a sensor [1].

The differential biopotential readout circuits amplify biosignals sensed on two electrode terminals, filter them to reject interference, and convert the amplified signal to the digital domain where it is further processed and transmitted wirelessly to an external device [2]. The specifications for readout blocks are very tough. Besides typical limitations present in all implantable devices (miniature size and very low power consumption), readout circuits need to handle very weak, low-frequency signals in the presence of strong interference and noise. Furthermore, offsets originated from the electrodes can be much higher than the biosignal's amplitude, and therefore, they need to be carefully removed in order to avoid saturating the amplifiers.

The electrochemical biosensing readout circuits are implemented with amplifier configurations around three-electrode transducers, shown in Figure 1.2. Each transducer is composed of a working electrode (WE), a reference electrode (RE), and a counter electrode (CE). An electrochemical reaction takes place on the WE when a particular DC potential is present on the RE. The electrodes can be functionalized with enzymes so that they have a selective action on a particular target substance [3]. As the RE is only used to sense the potential, the CE should source/sink the current needed to close the loop. In electrochemical biosensing, the readout circuits should

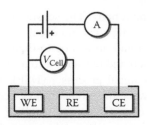

FIGURE 1.2   Three-electrode transducer.

set the RE potential to a precise value and measure the current flowing through the WE. These circuits have to deal with multiple challenges. The WE's current can vary from nanoamperes to microamperes [4]; therefore, the amplifiers should be able to handle a very large dynamic range. This is only possible by fulfilling stringent noise requirements. The connection of the WE and RE is also a critical issue, since both positive and negative cell voltages may be needed to bias a chemical reaction (in general, negative voltages are difficult to generate from a single power supply).

To understand the challenges present in implantable readout circuit design, the Sections 1.2.1 through 1.2.3 will review the electrical characteristics of the biological signal source within the body, namely, the biopotential signals and the electrode impedances. The latest biopotential readout AFE circuit solutions will be introduced, and their advantages/disadvantages will be discussed. Additionally, the latest trends in electrochemical biosensing readout AFE circuits will be presented. Finally, the latest advances in ultra-low-power ADCs for implantable devices will be shown.

1.2.1   Biopotential Sensing

1.2.1.1   Biopotential Signals

Typical sensors detect a particular characteristic in the environment and transform it into an electrical voltage or current. Biopotential sensors are not so different to typical sensors. The information in biosignals is generated by the electrochemical reactions within living cells. Accordingly, the biosignal activity is already available in the electric domain. However, voltages and currents are produced by ion exchanges instead of the flow of electrons. The dominant ions that are responsible for these electrochemical reactions are potassium (K^+), sodium (Na^+), and chloride (Cl^-). At the cell level, the cell's membrane presents some degree of selectivity for the kind of ions that can leave its wall, being more permeable to K^+ ions than to Na^+ ions. The K^+ ions leaving the cell create a diffusion gradient that makes the interior of the cell's membrane more electrically negative than the exterior. Electrostatic equilibrium is achieved at a particular electric field, and if no external disturbance is applied (a condition called resting potential), the voltage drop at the membrane's cell settles at around −70 mV. If an electrical stimulation is applied, the permeability of the membrane changes and allows more Na^+ ions to enter the cell. The diffusion gradient changes, the potential increases, and its sign can eventually change. The potential increase could reach a maximum voltage of around +40 mV where the permeability to Na^+ starts to decrease, the permeability to K^+ increases, and the voltage returns to the resting potential. Consequently, electrical stimulations on living cells produce voltage waveforms that are related to its physiological conditions.

In nature, electrical stimulations originate from the central nervous system. The biopotentials that these stimulations produce on the cells reflect many physiological conditions that are useful in diagnosis, prevention, and treatment of medical diseases. Electrical waveforms that are sensed at different places on the chest, for instance, provide useful information about the heart's activity. This is the basis of electrocardiography. Likewise, brain activity can be monitored by sensing different points on the scalp, through electroencephalography, and skeletal muscle electrical activity can be recorded through electromyography. As mentioned earlier, the electrical variations are the result of ion exchanges in the cells. Accordingly, transducers are required to transform ionic currents and voltages into electronic currents and voltages so that they can be processed by analog circuits. These transducers are the biopotential electrodes.

1.2.1.2 Biopotential Electrodes

Physiological substances function mostly as electrolytes; their electric properties are uniquely defined by ionic charges since they do not have free electrons. Electrodes, on the other hand, do not have ions, but let electrons flow freely. The objective of the biopotential electrodes is to transform ions to electrons and vice versa by means of reduction and oxidation chemical reactions (redox) at the electrode interface. This electrochemical process creates a voltage drop at the electrode interface, the so-called half-cell potential. This potential is important because voltages are sensed differentially by using two electrodes, and mismatches result in a DC offset of typically several millivolts. Figure 1.3 shows the general electrical model of the bioelectrode, where V_{hc} represents the half-cell potential, R_S is the electrolyte resistance, and C_A and R_A are the capacitance and resistance at the electrode–electrolyte interface, respectively. The values of these components depend on the electrode type, which can be polarizable or nonpolarizable. Polarizable electrodes are dry electrodes that touch the tissue directly without any substance, which helps improve the contact properties. They behave mostly as leaky capacitors and present very high impedances at low frequencies; therefore, extremely high input impedances are required for the readout circuits. When a helping substance, such as gel, is used to improve the contact properties (wet electrodes), the electrode's impedance is mainly resistive. These electrodes are called nonpolarizable and have the advantage that they present a much lower impedance than the dry electrodes. This advantage is shadowed by the need to apply gel, which may be impractical in many applications (for instance, in implants), and therefore dry electrodes may be the only alternative. Latest

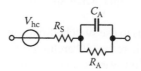

FIGURE 1.3 Electrical model of the electrode.

improvements show that dry electrodes' impedance can be substantially reduced by using micromachined microelectromechanical systems (MEMS) spikes [2].

1.2.1.3 Biopotential Amplifiers

Biopotential signals are typically in the order of microvolts and need to be amplified to at least several hundreds of millivolts before they can be converted to the digital domain. The amplifying chain is normally implemented by cascading a biopotential amplifier with variable gain amplifying stages. The design of biopotential amplifiers is challenging since undesired signals such as DC offsets (coming from the electrodes and from circuit mismatches), low-frequency noise, and electromagnetic electrostatic interference can be orders of magnitude larger than the biopotential levels and therefore need to be completely filtered or strongly suppressed. All of these requirements have to be accomplished while consuming as little power as possible.

DC offsets that originated from the half-cell potential mismatches on the electrodes can reach tens of millivolts. They must be removed or strongly suppressed since they can easily saturate the amplifier. Accordingly, some means of high-pass filtering (HPF) is mandatory. The implementation of an HPF at such low frequencies requires large capacitors that often cannot be integrated and must be placed as off-chip components.

The low-frequency noise in complementary metal-oxide semiconductor (CMOS) devices is dominated by $1/f$ noise. In many cases, the signal-to-noise ratio is already very low because of thermal noise from the electrode's resistances. Therefore, more performance degradation owing to $1/f$ noise cannot be accepted. Typically, chopper techniques are used to mitigate $1/f$ noise and reduce the amplifier's offset at the expense of power consumption.

One of the main concerns when working at these low frequencies is the omnipresent 50/60 Hz AC power line capacitive coupling. The AC power line coupling can easily reach levels of tens of millivolts, and it appears as a common-mode signal at the input of the electrodes. A very high common-mode rejection ratio (CMRR) is required in order to avoid this interference to appear at the input of the ADC. Furthermore, the electrode impedances are quite large and prone to mismatches. As a result, the voltage divider at the input of the amplifier creates a common mode-to-differential conversion mechanism. A very high input impedance is required in order to mitigate this mechanism.

The three-opamp instrumentation amplifier (IA) shown in Figure 1.4 has been the classical solution for discrete biopotential amplifier implementations. Fully integrated CMOS implementations of this IA are possible; however, process variations on the resistors are the cause of mismatches in the feedback networks and therefore incomplete common-mode cancellation. Even small mismatches make it very difficult to achieve the extremely high CMRR specifications without requiring complex calibration solutions.

A more interesting option is to use switched-capacitor (SC) amplifiers. The advantage of this approach is that there are SC architectures that are intrinsically insensitive to $1/f$ noise and offsets. However, in very sensitive applications, the specific kT/C noise in SC circuits could be a bottleneck since large capacitor values are required. Driving these large capacitors is not a trivial task. Good performance requires fast

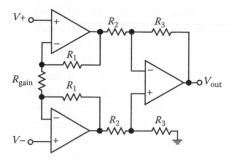

FIGURE 1.4 Three-opamp IA.

amplifiers so that voltages settle quickly in the capacitors. This is achieved at the expense of high power consumption.

A very popular biopotential amplifier topology is the current feedback or current balancing IA (CBIA), which consists of a transconductance input stage followed by a transimpedance output stage [5], as it is shown in Figure 1.5. The main advantage of this topology is that both the transconductance and transimpedance stages require a single resistor instead of resistor pairs that should match perfectly as in the classical IA. Therefore, the mismatch problem that affects the performance of the previous topologies disappears. Moreover, the CBIA topology is more power efficient than the IA since it should drive only two resistors instead of seven. The noise at low frequencies, however, remains the main issue.

Common circuit techniques to suppress the low-frequency noise and biopotential amplifier–related offsets consist of large-signal excitation (LSE), correlated double sampling (CDS)/auto-zero, and chopping modulation [6]. LSE consists of turning on and off the devices frequently so that the $1/f$ noise memory mechanism is reset. Its application is found mainly in SC circuits. CDS/auto-zero consists of estimating the offset and subtracting it from the signal [7]. The chopping technique involves

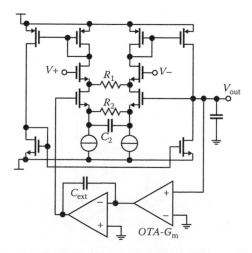

FIGURE 1.5 Current feedback IA.

modulation of the $1/f$ noise above the signal band. In practice, a combination of chopping and auto-zero techniques is needed in order to handle both offset and noise. In some cases, the designs implement coarse and fine auto-zero loops to handle different kinds of offsets. Nowadays, chopping and auto-zero techniques are extensively applied in biopotential amplifiers. Figure 1.6a shows the conceptual diagram of a chopper amplifier. This configuration cannot be directly applied to biopotential amplifiers since the electrode's offset can saturate the output. A solution to this problem is shown in Figure 1.6b where an HPF is implemented by using a servo-loop. Similarly, another servo-loop can be implemented in order to suppress the amplifier's DC offsets, as it is shown in Figure 1.6c. The previous techniques are very powerful. Nevertheless, they come with a high price tag: chopping entails up-conversion and amplification of the biosignals at frequencies much higher than their bandwidth. The required performance in the amplifiers can only be achieved by increasing the power consumption.

Reduction of the power consumption is a major driving force in implantable biopotential sensing. The introduction of floating-gate transistors in the biasing circuits eliminates the need of current mirrors and therefore reduces the power consumption substantially [8]. This technique in combination with pure capacitive feedback [9] can provide a good trade-off between performance and power consumption. In addition, the use of capacitors at the input of the amplifier has the advantage that it makes

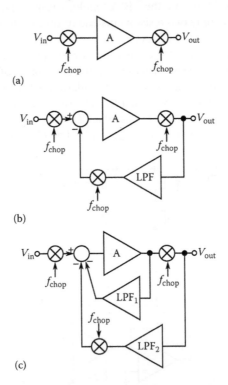

FIGURE 1.6 Chopper amplifier configurations: (a) chopping concept, (b) servo-loop to implement HPF, and (c) additional servo-loop to suppress amplifier's offset.

the readout system intrinsically safe as no DC currents other than leakage currents can flow between the body and the readout's electronics.

Further improvements can be achieved by combining the advantages of both continuous-time (CT) and SC circuit techniques. For instance, the CT capacitive coupled IA can be followed by a passive SC high-pass filter [10]. This is a very flexible solution since the filter characteristics are easily programmable.

1.2.2 ELECTROCHEMICAL BIOSENSING

The readout circuits for electrochemical biosensors process the output currents from three-electrode transducers. Most of the circuit implementations transform the current signal to a voltage signal by using a transimpedance amplifier. The resulting voltage signal is converted to the digital domain by using an ADC and further processed by digital circuits. Nevertheless, the A/D conversion and digital processing can be very expensive in power terms. Therefore, it has become more common to use simple implementations of current–frequency converters. Practical implementations of electrochemical biosensors using this technique have shown that ultra-low-power solutions in the order of just a few microwatts are possible [11]. This is important because it enables the possibility of implementing miniature-size biosensors that can be powered remotely, for instance, by using inductive coupling [12].

The critical challenges involved in the design of CMOS readout circuits for electrochemical sensors are basically the same as in biopotential readouts. Low-frequency noise, stringent power requirements, and interference remain the most important considerations during circuit design. The same circuit techniques that are employed in biopotential readout circuits and that were discussed in Section 1.2.1 are used in electrochemical sensor circuits. For instance, SC and chopper stabilization techniques have been recently used in order to improve the noise performance of electrochemical sensors [13]. Since the electrochemical reaction is a rather slow process, there is almost no substantial power increase when applying chopper modulation.

The low-power characteristics present in the latest electrochemical readout circuits show the potential of integrating different biochemical sensors in the same implantable device. Moreover, fully integrated biosensor arrays by using microfabrication and electrodeposition have been recently demonstrated [14]. This is a major driving force for the development of application-specific integrated circuit (ASIC) solutions, which are able to handle multiple biosensors. One of the latest advances in this field is a fully integrated readout circuit, which multiplexes different electrochemical biosensors and provides both cyclic voltammetry and chronoamperometry measurements as it is shown in Figure 1.7 [15]. These kinds of ASIC solutions, which consume power in the sub-milliwatt range, will eventually enable miniaturized implantable biochemical sensors that are expected to revolutionize medical diagnoses and treatment techniques in the near future.

1.2.3 ADCs FOR IMPLANTABLE READOUTS

The ADC, which provides the link between the analog signal at the output of the AFE readout and the DSP, is a key building block in implantable readouts. The

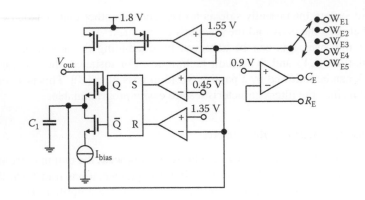

FIGURE 1.7 Amperometer with current–frequency conversion [15].

requirements of the ADC depend on several aspects, for example, the characteristics of the biopotential signals, the performance of the AFE readout, and the system architecture. The frequency content of different biopotential signals ranges from near DC to 10 kHz, and the amplitude range is from a few microvolts to hundreds of millivolts [16]. Since biosensors are usually attached to or implanted into human bodies and powered by either portable batteries or harvested energy, ultra low power and packaging in a small form factor are often required. These constraints, complemented by the implantable biosensors' requirements, impose challenges in designing power- and area-efficient ADCs that meet both the bandwidth and resolution specifications.

There are two common approaches to process biopotential signals, based on either a medium-resolution Nyquist ADC, such as a successive approximation register (SAR), or a high-resolution oversampling ADC, such as a Sigma-Delta ($\Sigma\Delta$) [17]. As shown in Figure 1.8a, a medium-resolution ADC requires a voltage gain amplifier (VGA) following the low-noise IA to adapt the signal amplitude to the ADC's input dynamic range. In addition, a Nyquist ADC is often preceded by a high-order anti-aliasing filter (AAF), with a fixed cutoff frequency that limits the system bandwidth.

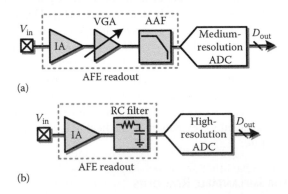

FIGURE 1.8 Simplified block diagram of biopotential acquisition systems based on (a) a medium-resolution Nyquist ADC and (b) a high-resolution oversampling ADC.

By employing a high-resolution oversampling ADC, as shown in Figure 1.8b, the requirements of the AFE readout can be much relaxed: the VGA can be eliminated and the high-order active AAF can be replaced by a simple RC filter. The significant reduction of AFE complexity potentially implies both smaller chip area and lower power consumption. On the other hand, it is traded with the challenge of designing a high-resolution ADC with high power efficiency.

The choice of ADC architecture depends on the bandwidth, resolution, and power requirements of different biosensor applications. Among existing ADC architectures, the SAR ADC has emerged as the most suitable solution for ultra-low-power medium-resolution applications. For low-power high-resolution applications, the $\Sigma\Delta$ ADC is the most power efficient. When it comes to the application scenario that requires multiplexing and digitizing biopotential signals from electrode arrays, the incremental $\Sigma\Delta$ ($I\Sigma\Delta$) ADC becomes necessary. The remainder of this section reviews state-of-the-art CMOS ADCs dedicated to various implantable readouts.

The SAR ADC employs a "binary-search" algorithm to convert the input sample in a "one-bit-per-cycle" fashion. Because of its minimum analog complexity (usually only a comparator), the SAR ADC is highly power efficient and has been widely adopted in state-of-the-art biosensor systems. In the work of Zhang et al. [18], a 53-nW SAR ADC for medical implants is presented. A dual-supply voltage scheme is used to allow the SAR logic to operate at 0.4 V, reducing the overall power consumption of the ADC without performance loss. In the study by Tang et al. [19], a SAR ADC with scalable sampling rate and reconfigurable resolution is proposed for an implantable retinal prosthesis. Using a charge folding method, the ADC has an input range of 1.8 V, while using a supply voltage of only 0.9 V. A single-ended SAR ADC suitable for multichannel neural recording is presented in Tao and Lian's work [20]. Because of several practical issues, it is challenging for the SAR ADC to achieve high resolution (i.e., more than 14 bits), while still maintaining high power and area efficiency. Therefore, more research efforts and explorations are needed for designing power- and area-efficient SAR ADCs for high-resolution biosensor applications.

The $\Sigma\Delta$ ADC employs oversampling and noise-shaping techniques. It achieves high resolution with relaxed component matching requirements compared to a Nyquist ADC. A switched-opamp $\Sigma\Delta$ ADC, which employs a single-phase scheme technique to improve the dynamic range and reduce the circuit complexity, is presented by Goes et al. [21]. The proposed ADC is capable of digitizing different biopotential signals only by reconfiguring the digital filter. In the work of Cannillo et al. [22], a multibit CT $\Sigma\Delta$ ADC used for biopotential acquisition systems is presented. By using a time-domain quantizer and a pulse-width modulated digital-to-analog converter (DAC), the proposed ADC alleviates the mismatch problem of multilevel DACs. One limitation of the conventional $\Sigma\Delta$ ADC is that it cannot be directly used in time-multiplexed environments.

The $I\Sigma\Delta$ ADC is structurally a $\Sigma\Delta$ ADC in which the $\Sigma\Delta$ modulator loop filter and digital filter are periodically reset. Therefore, the $I\Sigma\Delta$ ADC performs memoryless conversion, and a one-to-one mapping between the input and output samples is achieved as in Nyquist ADCs. As a result, $I\Sigma\Delta$ ADCs are well suited for multichannel applications, acting as high-resolution Nyquist-rate ADCs. In the study by

Agah et al. [23], an extended-range (ER) IΣΔ ADC is described, which is designed for handling time-multiplexed samples from bioluminescence sensor arrays. In the proposed ER IΣΔ ADC, the residual error from a discrete-time IΣΔ modulator is encoded using a SAR ADC so as to improve the resolution. In Garcia et al.'s work [24], a third-order CT IΣΔ ADC targeting multichannel neuropotential sensors is presented.

In multichannel biopotential recording systems, the rapid-growing channel count may impose a higher speed requirement for the multiplexed ADC. To achieve an increased ADC conversion rate without compromising the resolution, the "pipelining" approach can be applied to IΣΔ ADCs. A pipelined CT IΣΔ ADC, shown in Figure 1.9, is proposed by Tao et al. [25]. By pipelining n stages of Lth-order CT IΣΔ ADC, the proposed architecture provides the design freedom coming from both the pipelined ADC and the ΣΔ ADC. In addition, CT implementation is employed to take advantage of the relaxed settling and bandwidth requirements of active blocks.

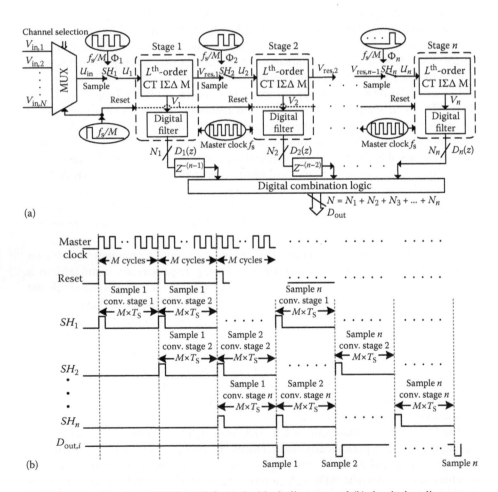

FIGURE 1.9 Pipelined CT IΣΔ ADC: (a) the block diagram and (b) the timing diagram.

The CT $I\Sigma\Delta$ modulator in each stage processes each sample at an oversampling rate of f_s. The number of clock cycles per conversion, M, is equivalent to the oversampling ratio in conventional $\Sigma\Delta$ ADCs. After M cycles, the residue of each stage, $V_{res,i=\{1,2,...n-1\}}$, is sampled and fed into the subsequent stage. At the same time, the digital filter produces a valid result, $D_{i=\{1,2,...n\}}$. The residue of each stage can be obtained at the Lth integrator's output, when the CT $\Sigma\Delta$ loop filter is implemented with a feed-forward topology [26]. After the sample has propagated through all the stages, the valid digital outputs, $D_1(z) ... D_n(z)$, are combined to yield the final digital word, D_{out}. In the end, the loop filters and the digital filters are reset and ready to accept the next sample. Because of pipelining operation, the ADC can achieve a resolution of $N = N_1 + N_2 + \cdots + N_n$ and its conversion rate is set by the time of a single stage, M/f_s. The theoretical resolution, N, can be estimated as [25]

$$N = \sum_{i=1}^{n} \log_2 \left\{ \left(\frac{U_{i,max}}{V_{FS}} \frac{(M+L-1)!}{L!(M-1)} + 1 \right) \left(a_1 \prod_{j=1}^{L} c_j \right) \right\}, \qquad (1.1)$$

where $U_{i,max}$ is the peak amplitude of the input signal, V_{FS} is the full-scale amplitude, and b_1, $c_{1...L}$ are the loop filter coefficients. Although each D_{out} comes with a latency of $n \times M/f_s$, this is tolerable in most of the biosensor applications. The proposed architecture provides inherent flexibility in achieving different conversion speeds and resolutions: the conversion rate can be tuned by changing M while maintaining f_s; different resolutions can be achieved by using different numbers of stages. Therefore, this architecture is capable of accommodating various biopotential signals with different bandwidth and dynamic range requirements.

The presented pipelined CT $I\Sigma\Delta$ ADC architecture has three design parameters. (i) the number of pipeline stages, n; (ii) the number of cycles per conversion, M; and (iii) the modulator order in each stage, L. In order to find an optimal combination of these design parameters, different configurations of (n, L) have been analyzed, simulated, and compared when identical stages were considered, as shown by Tao et al. [25]. It has been demonstrated that the configuration of $(n = 2, L = 2)$ is the most power- and area-efficient solution among many configurations, when targeting 14-bit resolution and 4-kHz signal bandwidth. An implementation test case of the proposed architecture, which pipelines two stages of second-order CT $I\Sigma\Delta$ ADCs [26], is shown in Figure 1.10. The test-case ADC, targeting neural recording systems, has been designed and fabricated in a standard 0.18-μm CMOS technology. Active-RC integrators are used to implement the loop filters of each second-order CT $I\Sigma\Delta$ modulator. For the targeted 14-bit resolution, $M = 40$ is selected, which corresponds to a 320-kHz clock frequency in the $\Sigma\Delta$ modulators. Measurement results show that the prototype ADC achieves a peak signal-to-noise-plus-distortion ratio of 75.9 dB and a dynamic range of 85.5 dB over a 4-kHz bandwidth, while consuming a static power of 34.8 μW. This corresponds to a state-of-the-art figure of merit of 0.85 pJ/conv.

In addition to these conventional ADC architectures based on uniform sampling, several ADC alternatives, which take advantage of the burst-like property of biopotential signals, have been recently proposed. One of the alternatives is level-crossing

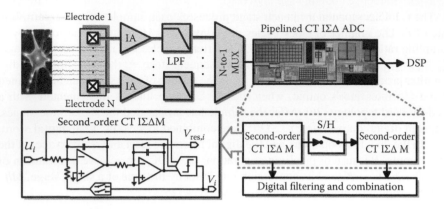

FIGURE 1.10   A pipelined CT $I\Sigma\Delta$ prototype ADC for neural recording systems.

sampling ADC [27], where samples are generated only when the input signal crosses the threshold levels. Therefore, for biopotential signals that are sparse in time domain, comprising both low-frequency and high-frequency contents, the level-crossing ADC generates fewer samples and thus consumes less power [28]. Another alternative, that is, a bypass window SAR ADC [29], achieves excellent energy efficiency on the order of femtojoules per conversion step when processing small-variation-in-amplitude biopotential signals. The ADC selects switching sequences to skip several conversion steps, when the signal is within a predefined bypass window. Consequently, the power consumption of the ADC is much lower than a conventional SAR ADC. The signal activity–based A/D conversion is a potential future trend for developing ultra-low-power ADCs for biosensor applications. On the other hand, advances in the ADC architectures and circuit techniques based on the conventional A/D conversion can also provide promising solutions in the future.

1.3 SELF-POWERING

Biomedical devices implanted in the human body require an electric power source to operate. Since, in these applications, it is unpractical to provide powering with cables, the typical solution has been based on primary batteries as part of the implant. This solution, however, has many drawbacks. First, batteries are bulky components that are difficult to be integrated in miniaturized implantable sensors. Second, the materials used in the fabrication of primary batteries are toxic and leakages can be harmful. Third, batteries have a limited life span; once they have depleted their charge, the batteries require to be replaced with the inconvenience of patients needing surgical intervention. So far, the lifetime of the implantable devices has been extended by using ultra-low-power circuit design techniques. Nevertheless, the usage of batteries remains one of the central problems in the development of implantable devices. The most promising solution to overcome the current challenges and obstacles presented by these conventional powering techniques is to use emerging energy harvesting technologies, which can enable a complete new generation of lifelong power-independent implantable systems. Implantable devices can harvest energy from different sources that originated not only inside the body but also around it, as shown in Figure 1.11. Reported research results [30] showed that around 4 μW/cm² can be harvested from mechanical movement such as walking. The body heat can also be used since thermal gradients are able to source approximately 20 μW/cm². In addition, the human body can be used as a source of electrochemical energy when biological substances are used as components in biofuel cells. For instance, a single glucose biofuel cell can provide approximately 190 μW/cm² [31]. Besides these internal sources that are associated to the human body, implantable devices can harvest external electromagnetic energy from radio-frequency (RF) transmitters located nearby. Achievable power levels for RF harvesters vary largely since this technique is very sensitive to link budget parameters such as frequency, distance, antenna size, gain, and so on. Harvested power values in the order of microwatts are typical numbers for this technique [32]. If the powering device is allowed to be in the close vicinity to the implant (several centimeters), it is possible to transfer power by using induced magnetic coupling [12]. Power levels in the order of milliwatts can be harvested from inductive coupling [33]. RF and inductive coupling techniques have the important advantage that can also be used to establish communication links between the implanted device and the external world.

The power level obtained at the energy harvester's output is usually unpredictable and varies over time as it depends on the environment. The power consumed by the biosensor could also vary over time since it depends on its mode of operation. Since the harvested energy is not always present or abundant, any portion of the harvested energy must be preserved so it can be used when needed. Therefore, as it can be seen in Figure 1.1, the energy harvester should be complemented by a temporary energy storage element, such as supercapacitors. Additionally, a power management module, which transfers the energy from micropower harvesters to biosensor circuitry, is required. The design of this module is quite challenging considering that it should provide high-efficiency energy transfer for very low input power and high voltage conversion ratios [34].

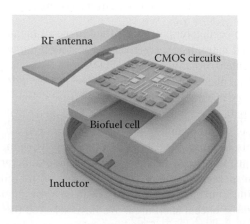

FIGURE 1.11   Self-powered implantable biosensor.

1.4   SUMMARY

In this chapter, we have introduced the basic principles of implantable biosensors with particular focus on readout circuits. Most contemporary implantable biosensor applications are challenged by power consumption, which affects the size because of the required batteries and the patient's comfort as a result of the materials used, the implantation procedure, and the need to replace the aging batteries. To minimize the size, increase energy autonomy, and make the implantable biosensors unobtrusive, tremendous efforts are made to improve the power efficiency and reduce the power consumption of electronic circuits. Additionally, there is an urgent need for new energy harvesting, storage, and processing methods that can provide the voltage and power required by the implantable biosensors and allow self-powering operation. Therefore, battery-free implantable biosensors must work under very limited power budget constraints and ultra-low-power solutions are required for all building blocks, from the readout circuitry to digital signal processing and wireless communications. The readout circuitry is critical as it provides the direct connection with the biological environment and it is strongly dependent on the application and the type of biosignal. The readout's power efficiency has been addressed by applying a power-aware design approach and ultra-low-power circuit techniques to implement the AFE and ADC. Self-powering of implantable devices is a very challenging task owing to the limited options for extracting electrical power within the human body. Emerging energy harvesting technologies within the human body, in combination with highly efficient storage and power management techniques, can enable a new generation of power-independent implantable biosensor systems.

REFERENCES

1. P. D'Orazio. 2003. Biosensors in clinical chemistry. *Clinica Chimica Acta*, vol. 334, no. 1–2, pp. 41–69, Aug.
2. R. F. Yazicioglu, C. Van Hoof, and R. Puers. 2009. *Biopotential Readout Circuits for Portable Acquisition Systems*. Springer Netherlands: Springer Science + Business Media B.V.
3. A. J. Bard and L. R. Faulkner. 2001. *Electrochemical Methods: Fundamentals and Applications, 2nd ed.* New York: Wiley.
4. M. M. Ahmadi, G. A. Jullien, and L. Fellow. 2009. Current-mirror-based potentiostats for three-electrode amperometric electrochemical sensors. *IEEE Trans. Circuits Syst. I, Reg. Papers*, vol. 56, no. 7, pp. 1339–1348.
5. M. S. J. Steyaert, W. M. C. Sansen, and A. Zhongyuan. 1987. A micropower low-noise monolithic instrumentation amplifier for medical purposes. *IEEE J. Solid-State Circuits*, vol. sc-22, no. 6, pp. 1163–1168.
6. C. Enz and G. Temes. 1996. Op-amp imperfections: Autozeroing, correlated double sampling, and chopper stabilization. *Proc. of the IEEE*, vol. 84, no. 1, pp. 1584–1614.
7. E. M. Spinelli, N. Martínez, M. A. Mayosky, and R. Pallàs-Areny. 2004. A novel fully differential biopotential amplifier with dc suppression. *IEEE Trans. Biomed. Engineering*, vol. 51, no. 8, pp. 1444–1448, Aug.
8. T. Wang, M. Lai, C. Twigg, and S. Peng. 2014. A fully reconfigurable low-noise biopotential sensing amplifier with 1.96 noise efficiency factor. *IEEE Trans. Biomed. Circuits Syst.*, vol. 8, no. 3, pp. 411–422.

9. F. Zhang, J. Holleman, and B. Otis. 2012. Design of ultra-low power biopotential ampli-fiers for biosignal acquisition applications. *IEEE Trans. Biomed. Circuits Syst.*, vol. 6, no. 4, pp. 344–355.

10. L. Yan, P. Harpe, M. Osawa, Y. Harada, K. Tamiya, C. Van Hoof, and R. F. Yazicioglu. 2014. A 680 nA fully integrated implantable ECG-acquisition IC with analog feature extraction. In *Proc. IEEE International Solid-State Circuits Conference (ISSCC)*, Feb., pp. 418–419, San Francisco, USA.

11. Y. Liao, H. Yao, and A. Lingley. 2012. A 3 μW CMOS glucose sensor for wireless contact-lens tear glucose monitoring. *IEEE J. Solid-State Circuits*, vol. 47, no. 1, pp. 335–344.

12. M. R. Haider, S. K. Islam, S. Mostafa, M. Zhang, and T. Oh. 2010. Low-power low-voltage current readout circuit for inductively powered implant system. *IEEE Trans. Biomed. Circuits Syst.*, vol. 4, no. 4, pp. 205–213, Aug.

13. H. M. Jafari and R. Genov. 2013. Chopper-stabilized bidirectional current acquisition circuits for electrochemical amperometric biosensors. *IEEE Trans. Circuits Syst. I*, vol. 60, no. 5, pp. 1149–1157, May.

14. A. Cavallini, C. Baj-Rossi, S. Ghoreishizadeh, G. De Micheli, and S. Carrara. 2012. Design, fabrication, and test of a sensor array for perspective biosensing in chronic pathologies. In *Proc. IEEE Biomedical Circuits & Systems Conference (BIOCAS)*, pp. 124–127.

15. S. Ghoreishizadeh, C. Baj-Rossi, A. Cavallini, S. Carrara, and G. De Micheli. 2014. An integrated control and readout circuit for implantable multi-target electrochemical biosensing. *IEEE Trans. Biomed. Circuits Syst.*, vol. 8, no. 6, pp. 1–8.

16. J. Webster. 2009. *Medical Instrumentation Application and Design*. New York: Wiley.

17. K. Soundarapandian and M. Berarducci. 2009, 2010. Analog front-end design for ECG systems using delta-sigma ADCs. *Texas Instruments, Tech. Rep.*, pp. 1–10.

18. D. Zhang, A. Bhide, and A. Alvandpour. 2012. A 53-nW 9.1-ENOB 1-kS/s SAR ADC in 0.13-μm CMOS for medical implant devices. *IEEE J. Solid-State Circuits*, vol. 47, no. 7, pp. 1585–1593.

19. H. Tang, Z. C. Sun, K. W. R. Chew, and L. Siek. 2014. A 1.33 μW 8.02-ENOB 100 kS/s successive approximation ADC with supply reduction technique for implantable retinal prosthesis. *IEEE Trans. Biomed. Circuits Syst.*, vol. 8, no. 6, pp. 844–856.

20. Y. Tao and Y. Lian. 2015. A 0.8-V, 1-MS/s, 10-bit SAR ADC for multi-channel neural recording. *IEEE Trans. Circuits Syst. I, Reg. Papers*, vol. 62, no. 2, pp. 366–375.

21. J. Goes, N. Paulino, H. Pinto, R. Monteiro, B. Vaz, and a. S. Garção. 2005. Low-power low-voltage CMOS A/D sigma-delta modulator for bio-potential signals driven by a single-phase scheme. *IEEE Trans. Circuits Syst. I, Reg. Papers*, vol. 52, no. 12, pp. 2595–2604.

22. F. Cannillo, E. Prefasi, L. Hernández, E. Pun, F. Yazicioglu, and C. Hoof. 2011. 1.4 V 13 μW 83dB DR CT-ΣΔ modulator with dual-slope quantizer and PWM DAC for biopotential signal acquisition. In *European Solid-State Circuits Conf.*, pp. 267–270, Helsinki, Finland.

23. A. Agah, K. Vleugels, P. Griffin, M. Ronaghi, J. D. Plummer, and B. A. Wooley. 2010. A high-resolution low-power incremental ADC with extended range for biosensor arrays. *IEEE J. Solid-State Circuits*, vol. 45, no. 6, pp. 1099–1110.

24. J. Garcia, S. Rodriguez, and A. Rusu. 2013. A low-power CT incremental 3rd order sigma-delta ADC for biosensor applications. *IEEE Trans. Circuits Syst. I, Reg. Papers*, vol. 60, no. 1, pp. 25–36.

25. S. Tao, J. Chi, and A. Rusu. 2015. Design considerations for pipelined continuous-time incremental sigma-delta ADCs. In *Proc. IEEE International Symposium on Circuits and Systems (ISCAS)*, pp. 1014–1017, Lisbon, Portugal.

26. S. Tao and A. Rusu. 2015. A power-efficient continuous-time incremental sigma-delta ADC for neural recording systems. *IEEE Trans. Circuits Syst. I Reg. Papers*, vol. 62, no. 6, pp. 1489–1498.
27. N. Sayiner, H. V. Sorensen, and T. R. Viswanathan. 1996. A level-crossing sampling scheme for A/D conversion. *IEEE Trans. Circuits Syst. II: Analog Digit. Signal Processing*, vol. 43, no. 4, pp. 335–339.
28. Y. Li, D. Zhao, and W. A. Serdijn. 2013. A sub-microwatt asynchronous level-crossing ADC for biomedical applications. *IEEE Trans. Biomed. Circuits Syst.*, vol. 7, no. 2, pp. 149–157, Apr.
29. G. Y. Huang, S. J. Chang, C. C. Liu, and Y. Z. Lin. 2012. A 1-μW 10-bit 200-kS/s SAR ADC with a bypass window for biomedical applications. *IEEE J. Solid-State Circuits*, vol. 47, no. 11, pp. 2783–2795.
30. R. Vullers, R. van Schaijk, I. Doms, C. Van Hoof, and R. Mertens. 2009. Micropower energy harvesting. *Solid-State Electron.*, vol. 53, no. 7, pp. 684–693, July.
31. A. Zebda, S. Cosnier, J.-P. Alcaraz, M. Holzinger, A. Le Goff, C. Gondran, F. Boucher, F. Giroud, K. Gorgy, H. Lamraoui, and P. Cinquin. 2013. Single glucose biofuel cells implanted in rats power electronic devices. *Sci. Rep.*, vol. 3, p. 1516, Jan.
32. T. Le, K. Mayaram, and T. Fiez. 2008. Efficient far-field radio frequency energy harvesting for passively powered sensor networks. *IEEE J. Solid-State Circuits*, vol. 43, no. 5, pp. 1287–1302, May.
33. J. Olivo, S. Carrara, and G. De Micheli. 2011. Energy harvesting and remote powering for implantable biosensors. *IEEE Sens. J.*, vol. 11, no. 7, pp. 1573–1586, July.
34. J. Katic, S. Rodriguez, and A. Rusu. 2014. Analysis of dead time losses in energy harvesting boost converters for implantable biosensors. In *Proc. IEEE International Conference NORCHIP*, pp. 1–4, Tampere, Finland.

2 Low-Temperature Microassembly Methods and Integration Techniques for Biomedical Applications

Serguei Stoukatch

CONTENTS

2.1   INTRODUCTION

Microassembly and packaging in a broad sense is a technique that interconnects microelectronics into a system level to form a functional product for the end user. Microassembly and packaging provide electrical and mechanical interconnection between microelectronics and the package. From different sources (Harper 2005), readers can encounter other names for such technique; often, it is referred to as semiconductors packaging, integrated circuit (IC) assembly or microassembly, interconnection technique, or just assembly, packaging, and so on. Further in the chapter, I am using the terms *microassembly* and *packaging* together as a combination of two words or separately as synonyms to express the same meaning as "microassembly and packaging."

By *microelectronics*, I meant electronic and microelectronic devices; electronic components such as ICs, microsystems (MS) as commonly known in Europe, microelectromechanical systems (MEMS, which is the typical name for MS in the United States), or micromachines in Japan; and discrete surface-mounted devices (SMDs) such as semiconductors resistors, capacitors, and so on. The second-level board assembly such as micro ball grid array (µBGA), ball grid array (BGA), chip-scale package (CSP), and so on often also falls into this definition. From the processing point of view, the microassembly already starts on the wafer level. Nowadays, the wafer is typically thinned down from its original thickness of 750 µm (for the 12″ wafer) or 450 µm (for the 8″ wafer) to a thickness of 200–300 µm or even thinner. Then, the wafer is singulated on individual dies typically by the sawing technique followed by the die mounting process. The die mounting process includes the die pickup process, die bonding to the package, and permanent die fixation by means of adhesives. The process ends with adhesive curing, which is typically performed at elevated temperatures, followed by wire bonding. Wire bonding provides the electrical interconnection between the die and the substrate. Then, the die is encapsulated in order to protect the assembly from mechanical damage and chemical effects from the environment. The most harmful effect comes from moisture, microparticles, corrosive gases, and so on, which are constantly present in the atmosphere. An alternative technology for wire bonding is flip-chip assembly, which provides simultaneous mechanical and electrical contact between the die and the package. Afterward, the package with the assembled dies is singulated on the individual assembly that forms the electronics components.

Recently, the definition for microassembly and packaging has gradually extended from IC and MS packaging toward assembly and integration of biochips, biomedical devices that often use microfluidic components, and wireless medical systems.

2.2   CHALLENGES AND TECHNOLOGY DRIVES

There are several challenges for modern microassembly and packaging. The most crucial ones are to enable environmental friendly technology (green and greener technology), continuous cost reduction demand, miniaturization (weight and size), time to market, performance, high-speed signal transmission, heat dissipation, and reliability. Such challenges can be addressed through a different set of actions. The actions are based on a smart system design and analytic modeling, developing and introducing new environmentally friendly materials, and technology that enables low-temperature processing. Often the introduction of new materials and processes might increase cost and final device performance and reliability; hence, a smart balance must be maintained throughout the device's life cycle.

2.3 IC, MS, AND BIOCHIP PACKAGING

The current trend in IC and MS packaging is to perform processing at low temperatures and to reduce the overall exposure time of thermal treatment during processing. There are several reasons to reduce overall exposure to thermal treatment. Generally speaking, excessive thermal treatment causes unwanted and often high stresses in the final assembly, typically called thermal-induced stresses. They might cause irreversible changes in material properties that could result in partial or full loss of the device's functionality.

The standard semiconductor packaging that was initially developed for standard IC CMOS (complementary metal-oxide semiconductor) assembly typically uses a high processing temperature; for example, a typical epoxy used for die mounting cures at temperatures of 150°C–175°C for 0.5–2 h (Licari and Swanson 2011). Wire bonding is a process that electrically interconnects the IC with the package and is typically performed at elevated temperatures as high as 150°C–220°C (Harman 2010). Another interconnection technique is the flip-chip assembly (Lu and Wong 2009), which requires 250°C–260°C for 3–5 min for solder reflowing in the case of lead-free soldering and up to 300°C–400°C for the thermocompressing process for gold stud ball bumping technology. Typically, molding or glob-topping and encapsulation requires 150°C–180°C for 1–3 h for curing and postmolding cure. For standard IC CMOS devices, such high-temperature thermal treatment is acceptable, hence generating substantial stresses in the final assembly and reducing the device's life span, which, in turn, might result in elevated failure rates. Using a smart design, thermomechanical finite element modeling (FEM) simulation, matching assembly materials such as IC, substrate, and die attach adhesive and using underfill on one side may keep thermal-induced stresses at an acceptable level. On the other hand, during the last 10 years or so, instead of using classic assembly materials, material suppliers are slowly introducing lower-temperature processing materials that limit the maximum processing temperature for the packaging.

Contrary to IC CMOS packaging, MS packaging (Gilleo 2005) drives the maximum processing temperature below 150°C and requires substantial reduction of the thermal treatment. Typically, thermal treatment that exceeds 150°C causes serious and irreversible changes in sensitive MS structures such as RF bridges and membranes and degrades their performance.

2.4 PACKAGING OF BIOCHIPS

The processing requirements change drastically for microassembly and packaging of biochips.

First, in this chapter, I will define what a biochip is and the implications it has on packaging. The most common and general definition for a biochip is "a hypothetical computer logic circuit or storage device in which the physical or chemical properties of large biological molecules (as proteins) are used to process information" (*Merriam-Webster Dictionary*).

Here, I define the biochip as a chip with biological functions that specifically performs biological reactions. The biochip can be just the active system that consists of typically a solid substrate with an active biofunctional layer, or it can be a fully functional or fully integrated and even an autonomous system. In the case of the fully integrated biochip, it often also has, in addition to the substrate with an active biofunctional layer, an integrated microfluidic device that supplies the test sample to the sensing area of the biochip. The biochip in turn performs a biological reaction and converts the biosignal, as a result of a biological reaction, into a physical signal. Physical signal examples include an electrical signal or an optical signal that processes using a readout module. In this case, the biochip becomes a biosystem or a biodevice. The biodevice in some literature is known as a lab-on-chip.

The biofunctional layer is typically obtained by growing or depositing biomolecules on the sensing area of the biochips. Examples of such biomolecules are deoxyribonucleic acid (DNA), proteins, and so on. Biofunctionalization is typically a complex and multistep process, and often before deposition of the biomolecules, the surface is treated with biocompatible polymers such polyethylene glycol (PEG). In this case, PEG acts as a surfactant. Additionally, PEG has many others biological applications: it acts as a precipitant for DNA isolation and protein crystallization, it is used to concentrate viruses, and so on. The biofunctionalization is often performed on the wafer level as a cost-effective and high-throughput process. In this case, after biofunctionalization, the wafer must be separated on the individual biochips. Then, the individual biochip must be integrated and become a part of a fully functional system. It is known that the standard packaging methods employ different physical and chemical processes or eventually a combination. Such processes and technology have the following features: high temperature, exposure to different chemicals, high pressure, ultrasonic energy, electromagnetic exposure of different nature, ultraviolet (UV) radiation, and so on. Such features can cause irreversible damage to the biofunctional layer, which dictates specific restrictions during processing. Also, it must have no direct contact with water (which typically happens during dicing) and dust (which is typically generated during the dicing process by sawing). The water flow comes under the pressure in direct contact with the biolayer, it may easily dissolve or just mechanically wash away the biolayer. The silicon dust generated during sawing may partially or fully cover the biolayer. The thermal exposure that takes place during die mounting, wire bonding, and encapsulation must be limited for all biomaterials and PEG. Specifically, the maximum temperature must not exceed 37°C; it is preferred that the temperature is within the 20°C–25°C range.

2.5 MATERIAL CHALLENGES

2.5.1 SUBSTRATES

It is obvious that not all materials are suitable to serve as a base material for electronic, MS, and biosystem packaging. The materials must have a combination of thermomechanical, electrical, optical, chemical, and biological properties. All such properties must be balanced to match the required specifications. Here, the most common properties and considerations are listed.

The mechanical properties are flexibility, tensile strength, and dimensional stability; the thermal properties are glass transition temperature (T_g), coefficient of thermal expansion (CTE), and operating temperature range. The electrical parameters are dielectric strength, dielectric constant, dielectric losses, and volume resistivity. The chemical considerations are moisture absorption and resistance to different chemicals. For biosystems, there is a specific requirement for biocompatibility. Recently, questions on the environmental impact of materials arise, which leads to the demand for green and recyclable materials. Last but not the least, there is cost consideration.

Another trend in IC and MEMS packaging is the use of cheaper materials, that is, substrates and materials with lower T_g. The substrate is usually referred to as the body of the base material for the package. The conventional nonorganic materials for substrates such as copper lead frames, other metals, ceramics, and glasses are being replaced or have already been replaced by organic materials such as FR4 (flame retardant, class 4) and bismaleimide triazine (BT) with T_g values of 150°C and 180°C, respectively. The organic FR4 and BT are cheaper compared to the nonorganic substrates. FR4 (woven glass–reinforced tetrafunctional epoxy materials) is so far the most common material for the printed circuit board (PCB). Currently, under the definition of advanced PCB materials, people understand non-FR4 substances as examples of materials that serve the same function as BT, which is widely used as a core material for μBGA and BGA types of packages. Compared to FR4 with a T_g of 150°C, BT has a higher T_g (180°C) and thus has better thermomechanical performance and has a higher cost.

Another material that was extensively studied and explored (Wojciechowski et al. 2001; Chandrasekhar et al. 2003; Ratchev et al. 2004) as a package base material and alternative to FR4 and BT is LCPs (liquid crystal polymers). LCPs are highly crystalline, inherently flame retardant, thermotropic (melt-orienting) thermoplastic materials. LCP is an extrudable and injection-moldable material. Unfilled polymers can be extruded for film or sheet, or melt spun for high-performance fibers or yarn. Current typical uses include electronic packaging, housing, and so on.

Reportedly, the LCP material trademark Zenite (DuPont registered trade names) is used as a body for the electronic package known as polymer stud grid array (PSGA). The PSGA is a registered trade name of Siemens Dematic AG. PSGA and related technology have been commercialized by Siemens Dematic AG. Similar to Zenite is Vectra developed by Celanese Corp. The main purpose for PSGA development was to have a cheaper, greener, and better alternative to BGA. They developed a technology to form a three-dimensional (3D) PSGA substrate by injection molding or hot stamping as an alternative to raw LCP. The PSGA package consists of a polymer injection-molded 3D body, the top side of which is flat and has a location for chip mounting and the back side has protuberances or polymer studs instead of BGA solder balls. The feedthrough holes are formed by laser drilling and then direct electroplated to provide the electrical contact between the top and the bottom of the PSGA. The PSGA body thickness is as thin as 350 μm, and the vias diameter is down to 100 μm. The high-density interconnect PSGA has conductive tracks as fine as

50 μm, and there is a 50-μm space between neighboring lines. The metallized studs, which are by nature lead-free, replace the solder balls within the BGA package. The package passed JEDEC moisture conditioning level 3 as well as standard reliability tests. PSGAs, similar to BGA and Quad Flat Pack (QFP), are surface-mount components used in a conventional surface-mount technology (SMT) assembly line without special tools and equipment needed.

Next to rigid base materials (Fletcher 2013), there is a group of flexible materials (Fjelstad 2007) that enables flex (often called flexible PCB) and rigid-flex circuit boards. The flex and rigid-flex circuit boards have extra functionality compared to the standard rigid PCB. They can fold, twist, wrap, and take different designated forms and shapes. In some sources, they are referred to as layers or flex layers because of their thickness (typically 0.025, 0.050, 0.100, and 0.200 mm), which is much thinner than that for FR4 and BT (0.4, 0.8, and 1.6 mm). Because of their thickness and ability to take different shapes, using them as a base material for mounting electronic components and MS has resulted in lighter and more compact packages. Currently, there are two main types of films that have been used in substrates for flexible PCB: polyester-based polyethylene terephthalate (PET) and polyimide (PI). PET films are the cheapest flex circuit board material used for microassembly and packaging and work well in nonhostile environmental conditions. They have limited electrical properties, and because of restricted thermomechanical characteristics, mainly lower T_g and higher CTE, there are limitations on manufacturability. PI films are used for critical applications where thermal, mechanical, and chemical resistance are key requirements. PI layers have superior electrical and thermomechanical material properties; hence, they are more expensive compared to FR4 and BT and much more expensive than (the least expensive) PET. There is another material for flexible PCB that belongs to the polyester family, which is polyethylene naphthalate (PEN). PEN in terms of performance and cost is a transition material between PET and PI. At a certain point in its development, it was intended as PET replacement. PEN is slightly better compared to PET in terms of performance. All polyesters cannot withstand a reflow soldering.

PETs are known on the market as Mylar and Melinex (DuPont registered trade names). Celanex is a trade name for thermoplastic polyester produced by Celanese. Teonex is the DuPont registered trade name for PEN films. Most of the PI films are made of Kapton (DuPont registered trade name).

Table 2.1 shows the most important thermomechanical properties of organic materials both rigid (FR4, BT, LCP, polycarbonate [PC], acrylonitrile butadiene styrene [ABS], and polylactic acid [PLA]) and flexible (PI, PET and PEN).

From a physical–chemical point of view, all plastic material can be grouped into two main categories: thermoset and thermoplastic material. The primary physical difference is that thermoplastics can be remelted back into a liquid, whereas thermoset plastics always remain in a permanent solid state. The thermoplastics can be melted above a specific temperature to a liquid and molded to a specified shape upon cooling. This process of heating and melting can usually be completed many times. Thermoplastic materials are typically easy to process and are definitely much easier to process than thermoset materials. There is a long list of such materials, but not all of them are suitable for electronic manufacture because of

TABLE 2.1

Packaging Material: Thermomechanical Properties

Material	T_g, °C	Melting Point, °C	Operating Temperature, °C	CTE, °C/ppm, α1/α2	E-Modulus, GPa	Moisture Absorption, % by Weight	Thermal Conductivity, W/mK
FR4	150	No	−55 to +125	14/70	24	3	0.25–0.35
PI	260	No	−80 to +210	20/40	2.5–5	2.5–3	0.12
BT	180	No	−55 to +140	13–14/14–17	32–35	0.2	0.35
LCP	120	300–400	260	15–18	8.5–17	0.02–0.1	0.2–0.3
PET	78–90	254	−60 to +105	27/n.a.	2.2–2.8	0.1–0.3	0.13–0.15
PEN	120	269	−60 to +150	18–20	3–5	0.1–0.3	0.15
PC	140	267	−40 to +130	25–70/n.a.	2.0–2.4	0.1–0.3	0.19–0.23
ABS	105	179	−20 to +80	56/n.a.	2.3	0.2–0.3	0.17–0.19
PLA	60–65	145–185°C	−20 to +40	30–60/n.a.	2.7–10	1–10	0.1–0.2

Note: n.a., not available.

their properties. In addition to having limited electrical properties, some thermoplastic materials have a very high melting point that often exceeds the solder reflow temperature. The use of other materials is too costly because of the excessive cost of raw material or demand for specific processing that is not compatible to the existing infrastructure.

In recent years, thermoset organic materials such FR4, BT, PI, and so on are being replaced by thermoplastic PET and PEN. In addition, other thermoplastic materials, such as LCP, PC, ABS, and PLA, are emerging in the electronic packaging market. The main reason for this is not cost but rather the reduction of electronic waste; all thermoplastic material can be recycled and theoretically reused unlimited amount of times. PET and PEN are not compatible with lead-free reflow soldering, and materials such as ABS and PLA will melt at temperatures even below the solder reflow temperature. To process such materials, an alternative technology must be used. PLA is one of the few biodegradable materials already used in electronics. An important application for both ABS and PLA is that they serve as materials for the filament of desktop 3D printers.

2.5.2 MATERIAL FOR BIOCHIPS AND MICROFLUIDIC SYSTEMS

In addition to the material requirement defined in Section 2.5.1, the materials for biochips and microfluidic systems must have specific biological and chemical properties. By biological properties, I primarily mean biocompatibility. Biocompatibility is typically defined as compatibility, that is, not causing any unwanted reaction (toxic, injurious, or any other kind) to biomaterials such as living tissue or living systems. There is an international standard (ISO 10993) to test the biocompatibility of a material. The material must be biocompatible to the levels required for the specific use. The standard ISO 10993 defines a device category, a contact regime, and a contact timescale.

The requirement for meeting specific chemical properties is typically understood as resistance to solvents used in life sciences applications.

The following is a list of biocompatible materials (CiDRA Precision Services: Life Sciences Polymer Materials; http://www.cidraprecisionservices.com) that are already widely used for biomedical applications such biochips, microfluidic devices, and so on: cyclic olefin copolymer (COC)/cyclic olefin polymer (COP), polyetheretherketone (PEEK), polymethyl methacrylate (PMMA), PC, and LCP. One can see that some of them (e.g., PC and LCP) are already known materials for electronic package; other materials such as COC/COP, PEEK, and PMMA have limited use in electronic packaging. Note that there are some other known biocompatible materials.

COC and COP are amorphous thermoplastic polymers. COP uses a single type of monomer during formulation. Both of them have the remarkable combination of high optical clarity, low water absorption, and good mechanical properties. Like many other thermoplastics, COCs and COPs can be extruded, injection molded, and machined. Depending on formulation, glass transition temperatures can range from 30°C to up 180°C. Because of their very good optical properties, COC and COP are used in microfluidic and microtiter plates. COCs and COPs are resistant to many solvents used in life sciences applications.

PEEK is a semicrystalline thermoplastic with very good thermal and mechanical resistance properties that are retained at high temperatures. Notably, PEEK is one of the few high-temperature polymers, with a service temperature of up to 250°C. Hence, it is an opaque material. PEEK can be injection molded or machined. It is resistant to almost all life sciences solvents except phenols. PEEK is used in fluidic fittings, pump and valve parts, and other high-wear and high-pressure fluidic components.

PMMA is a clear plastic that can be processed (by hot embossing), injection molded, and machined (mechanically or with the use of laser). PMMA can only be used at a limited temperature range of up to 85°C, and it is incompatible with most life sciences solvents. It is one of the key polymers used in low-cost microfluidics applications.

PC is another optically clear, hard plastic that can be processed via extrusion, injection molding, hot embossing, and machining. The characteristics of PC are similar to those of PMMA, but PC is stronger, has a wider service temperature range, and is more expensive. PC is resistant to only a few solvents used in life sciences applications. Because of its high impact resistance and optical transparency, it is used in windows in instrumentation and automation products. It is one of the key polymers used in low-cost microfluidics applications.

The following tables summarize the most important microfluidic application material properties (Table 2.2) and chemical compatibility to most life sciences solvents (Table 2.3) of polymeric thermoplastic materials such as COC/COP, PEEK, PMMA, PC, and LCP.

Aside from the materials mentioned in Tables 2.2 and 2.3, there are other life sciences polymer materials such as polyetherimide (PEI) (trade name, ULTEM), PET, polypropylene, polystyrene, and polysulfone. PEI is used in manifolds, pump and valve components, and some microfluidic and flow cell applications. The sheet form of PET is used in life sciences as a barrier material and in some flow cell and microfluidic applications despite the fact that it has relatively high water absorption characteristics. Metallization can be used to reduce water permeability. PET is resistant to most solvents used in life sciences applications. Polypropylene is commonly used for microtiter plates and reagent container fabrication. Polystyrene is still a common choice for microtiter plates, petri dishes, and test tubes used in benign applications.

2.6 INTERCONNECT AND PACKAGING MATERIALS

Interconnect and packaging materials follow the same trend as the packaging technology in general. They must be environmentally friendly and preferably recyclable, cheap, and low-temperature processable, and they must possess all necessary properties to enable high performance and good heat dissipation and to ensure high reliability.

In this chapter, I will briefly discuss the most common adhesives used for microassembly and packaging, including conductive and nonconductive adhesives, underfills, glob top, and encapsulants. Typically, both conductive and nonconductive adhesives are used for die attach, whereas nonconductive adhesives can also be used for molding, encapsulation, and glob top; in addition, nonconductive adhesives with extremely low viscosity are used in underfill applications.

TABLE 2.2
Thermoplastic Organic Material: Thermomechanical Material Properties

Material	T_g, °C	Melting Point, °C	Operating Temperature, °C	CTE, °C/ppm, α1/α2	E-Modulus, GPa	Moisture Absorption, % by Weight	Thermal Conductivity, W/mK
COC/ COP	30–180	190–230		60–70	1.7–3.2	<0.01 (COP) <0.1 (POP)	0.12–0.15
PEEK	157	332–338	220 to 249	13–32	6.2–23.2	0.06–0.18	0.22–0.94
PMMA	100–122	130	77 to 88	59–101	1.1–4.8	0.1–0.5	0.19–0.22
PC	140	267	−40 to +130	25–70/n.a.	2.0–2.4	0.1–0.3	0.19–0.23
LCP	120	300–400	260	15–18	8.5–17	0.02–0.1	0.2–0.3

Note: n.a., not available.

TABLE 2.3
Chemical Compatibility

Properties/Material	COC/COP	PEEK	PMMA	PC	LCP
Acetone	A	A	N	N	B
Acetonitrile	A	A	N	N	N
Dimethylformamide	N	A	N	N	N
Dimethylsulfoxide	A	B	N	N	N
Ethanol	A	A	N	A	A
Ethylene glycol	A	A	A	A	A
Glycerol	A	A	A	A	A
Isopropyl alcohol	A	A	B	B	B
Methanol	A	A	N	N	N
N-methyl-2-pyrrolidone	A	A	N	N	N
Phenol	N	N	N	N	N
Pyridine	A	A	N	N	B
Water	A	A	A	A	A

Note: Chemical rating key: A, little to no interaction; B, slight interaction; N, not compatible.

On the basis of the curing method, the adhesives used for microassembly can be classified into three main groups: heat cured, UV cured, and moisture curing adhesives.

The heat-cured adhesives can be divided into three main categories: conventional, fast cure, and snap cure types. The conventional adhesives cure at temperatures of 150°C–175°C for 0.5–2 h. The main challenge here is reducing the cure time (or the duration of the curing) or, in an ideal case, performing curing at room temperature if not lower. The fast-cure adhesives cure within 5–15 min at temperatures starting from 150°C. The manufacturers developed a so-called snap-curable die attach that requires typically only 1–5 min cure time at 150°C; there are some known adhesives that cure within seconds at 170°C–180°C.

There is a long list of heat cure adhesives currently available on the market. In addition to the three main categories, adhesives can be cured at reduced temperature but at a much longer time. Examples of such curing schedules are as follows: room temperature for 24 h, 40°C for 12 h, 60°C for 6 h, 80°C for 3 h, 120°C for 2 h, and so on.

Hence, the main challenge here is to achieve full curing at room temperature. Either the material requires very long curing time (several hours), or after such low temperatures, curing the material does not achieve the required properties.

A good alternative to heat curing adhesives are UV-cured adhesives that cure typically within 5–20 s at room temperature. The typical wavelength of the UV source is in the range of 250 to 500 nm, and the UV source intensity is 10–200 mW/cm². Hence, they are momentarily used for encapsulation. To cure them, UV light must access the material. In the case of the die attach application, the adhesive is covered by the substrate and the die, and UV adhesives are typically not used for that.

Moisture curing adhesives are very rarely used for electronic assembly mainly because of their long curing times (several hours) and lower functional properties compared with other adhesives.

2.7 TECHNOLOGY

Technology will play a crucial role in addressing challenges for modern micro-assembly and packaging. The technology must be environmentally friendly and should enable processing at low temperatures. The technology must contribute to such challenges through developing novel processes and techniques in handling new materials. Here, in this chapter, I will address these two main challenges.

Well-known and broadly introduced subtractive methods are gradually being replaced by additive methods (Suganuma 2014) that require less material, and hence, less waste is produced. For example, the conductive tracks on the bare substrate that serves as a body for the electronic package can be realized through a full additive process. A well-known example of this is conventional screen printing technology, which uses a conductive gold paste to print a pattern on ceramic substrates, where 100–150 µm is an absolute minimum value for line and space features. The main disadvantage of such process is the extremely high cure temperature for the conductive paste, which often exceeds 800°C. The inkjet printing technique can achieve features as small as 30 µm and does not require high cure temperatures. The ink jetting technique is developed and applied for low-T_g plastic substrates, both rigid and flexible.

2.8 AEROSOL JET PRINTING DEPOSITION

In the research conducted by Microsys lab, University of Liege, together with Sirris (Stoukatch et al. 2012), Aerosol Jet Printing (AJP) technology was evaluated and technology for microelectronic application was used. AJP is an innovative technology for a selective mask-less deposition of a wide range of materials (conductive, dielectric, biological, nanoparticles, etc.) at the micron scale, which is up to three times finer than the inkjet printing technique. Because it is a contactless technology, it suits a variety of applications, that is, flat and nonflat surfaces, flexible and rigid substrates, and complex 3D systems. The technology is particularly unique for deposition of a conductive silver paste on plastic substrates that have limited thermal budget in terms of the maximum permitted temperature and time of thermal exposure. The AJP silver layer has a very low sintering temperature (as low as 100°C–150°C), while the substrate remains at a lower temperature. The AJP silver layer fine pattern of 10 µm line and 10 µm space features is reportedly achieved.

I evaluated a set of different plastic substrates: FR4, PI, PC, and ABS, where the AJP silver pattern has been deposited using the Optomec AJ300CE system. Sirris, our research partner in the ongoing project, has developed the AJP silver deposition process. There are several commercially available silver inks that are suitable for AJP deposition; from among them, I selected the most common CSD-32 silver nanodispersion, manufactured by Cabot. The ink viscosity is 100 mPa·s. The CSD-32 silver ink is composed of engineered nanoparticles of 60-nm dimensions and 45%–50% weight in a liquid vehicle (glycol-based solvent).

During the deposition process, the ink is supplied by the system into pneumatic activation mode by dry nitrogen (carrier and sheath gas). The ink has been deposited on the test substrates and sequentially sintered by a laser. The laser with 100 mW power and 532 nm wavelength generated a 10-μm spot on freshly deposited silver layer. It requires only 100 μs dwell time to sinter the silver ink. Such short thermal exposure allows the thermal sintering on the silver nano-ink; meanwhile, the thermal-sensitive plastic substrates remain at low temperature. Sirris has developed a sintering process where only the deposited ink is exposed to elevated temperatures (approximately 100°C) and the thermal-sensitive plastic substrate remains at low temperature (below 80°C). The ink's manufacturer recommends performing sintering via the mass curing process, where the ink and the substrate are subjected to a temperature of 150°C for 30 min. Once the silver pattern is sintered, it is characterized using a 3D optical profilometer.

After the fabrication of the test samples, I assessed the suitability of the AJP deposited silver layer in terms of different conventional interconnect techniques used for electronic packaging; in this chapter, I will report only on the results of the evaluation for SMT. Each test substrate comprises a matrix of 30 SMD footprints suitable for 0402 components placement. For that, I used a conductive adhesive, which is the most common alternative to classic soldering SMT techniques. As a pick and place tool for mounting components, I used the automatic SMT pick and place system from Autotronik. The system also has a high-precision dispensing system and a bottom vision alignment system and can perform vision inspection before and after production. I investigated two curing schedules that differ by cure temperature and cure time: 3 h at 80°C and 15 min at 120°C. I placed SMD components on plastic substrates (Figures 2.1 and 2.2), cured at 80°C and 120°C for PI, FR4, and PC and at 80°C for ABS. The assembly has been inspected visually and evaluated by shear test, and sequentially, the failure mode for the shear test has been observed and characterized. The value for the shear test is in line with the results normally obtained on the rigid substrate with conventional finish and meets SMT specifications. The AJP deposited silver layer on all evaluated substrates (PI, FR4, PC, and ABS) is suitable for SMT using conductive adhesive, and PI and FR4 are suitable for lead-free soldering technique.

FIGURE 2.1   The SMD mounted on ABS with adhesive cured at 80°C.

FIGURE 2.2   The SMD mounted on PI with lead-free solder.

2.9   LOW-TEMPERATURE TECHNIQUE FOR THE ASSEMBLY OF A FULLY FUNCTIONAL AUTONOMOUS SYSTEM

The developed low-temperature technique was applied for the assembly of a fully functional prototype of Autonomous Micro-platform for Multi-sensors (AMM) developed by Microsys lab, ULg (Stoukatch et al. 2012).

To explore an alternative method to wire bonding in order to provide the electrical interconnection between the die and the flexible substrate, I used AJP silver ink. The silver track was processed by the AJP technique. In order to obtain a continuous conductive track between the terminals of the substrate and the bond pads on the die, the sample was prepared in a specific way. As depicted in Figure 2.3, a nonconductive fillet was deposited around the die bonded to the substrate.

The fillet has two main functions. First, the fillet is used to insulate the die sidewall to prevent short circuit between the die and the substrate. The second function is to ensure continuation of the AJP silver track. In principle, the AJP track can also be deposited on the vertical die sidewall; however, the sidewall then must be somehow electrically insulated. For the fillet material, I used UV curable die adhesive that cures at room temperature, on top of which the AJP conductive track was deposited, which starts on the substrate terminal and ends on the bond pad of the die.

The working prototype realized by AJP technology, where wire bonding was replaced by the silver conductive track, has a lower profile compared to the first prototype assembled using the conventional packaging technique. The AJP processed prototype is 0.5 mm thinner. Both prototypes have the same functionality. Figures 2.4 and 2.5 depict the interconnect area for both prototypes.

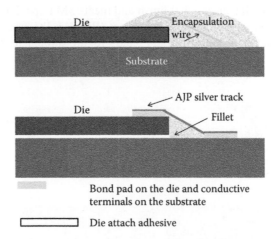

FIGURE 2.3   Top: Cross-sectional schematic view of the interconnection realized by conventional method using wire bonding and encapsulation technique. Bottom: Cross-sectional schematic view of the interconnection realized by the AJP silver layer.

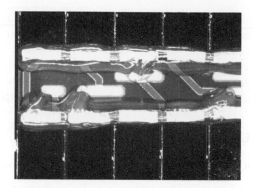

FIGURE 2.4 Image of the interconnect details between the substrate and the die, with the interconnection realized using conventional packaging methods (wire bonding and encapsulation).

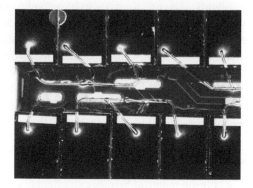

FIGURE 2.5 Image of the interconnect details between the substrate and the die, with the interconnection realized by the AJP silver layer.

2.10 ELECTRICAL CHARACTERIZATION

I performed extensive electrical characterization on the AJP conductive layers. For that, I used a four-point measurement method and performed electrical measurements on DC and AC (frequencies: 60 Hz, 1 kHz, 10 kHz, 100 kHz, 1 MHz, and 2 MHz). The test pattern was the AJP deposited silver track with the following features: 60 μm width, 3 μm thickness, and 20 mm length printed on the test substrates. At the 100-kHz frequency, the resistivity value remained constant. The experimentally obtained resistivity value on the test samples is 16 μOh·cm, which is 10 times higher than the resistivity value of bulk silver (1.6 μOh·cm); however, it is in line with the resistivity of thin film silver tracks applied by other deposition methods and low enough to support a variety of applications.

2.10.1 CONCLUSION

The AJP deposited silver layer on all evaluated substrates (PI, FR4, PC, and ABS) is suitable for SMT using the conductive adhesive; in addition, AJP on PI and FR4 is suitable for soldering technique.

The resistivity value is 16 µOh·cm, and it is similar to that of thin film silver conductive tracks applied by other deposition methods. The resistivity is low enough to support many applications.

It was also demonstrated that for specific applications (Autonomous Wireless Sensor Node prototype), conventional interconnection techniques such as wire bonding can be successfully replaced by AJP deposited conductive silver track.

2.11 MICROFLUIDIC DEVICES

2.11.1 CONVENTIONAL SEMICONDUCTORS MATERIAL AND TECHNOLOGY

The first generation of microfluidic devices similar to electronic devices was processed on silicon, glass, quartz, and even on ceramics and metals (Li and Zhou 2013). Such materials typically serve both as a substrate for the microfluidic device and as part of the microfluidic channel. Such technology at an early stage was very advanced for Si and glass processing and is widely used for CMOS and MEMS manufacturing. Typical examples of such technology include photolithography, deposition, etching, and so on. The main disadvantage of such technology is the high total cost of the devices, which is mainly attributed to the high cost of the raw material and the high fabrication cost. Despite this, a number of companies and research institutions still support this technology, and as a response, a variety of structural resists and related technology was introduced (Kakaç et al. 2010).

2.11.2 STRUCTURAL RESISTS

The structural resists or spin-on material brings extra benefits for the production of microfluidic devices on conventional semiconductor materials such as silicon and glass wafers. In this case, the microchannel is microfabricated in the structural resist; meanwhile, the entire device is processed on silicon or glass wafer. Most of them are well known in semiconductor manufacturing, are suitable for wafer-scale processing (up to 12″ silicone or glass wafers), can be processed by conventional lithography, often used for MEMS fabrication, and are above IC CMOS integration. Typical examples of such material are epoxy-based resists such as the SU-8 and KMPR series (Blanco Carballo et al. 2009), a variety of PI, and photoimageable and wafer-scale processable PDMS (polydimethylsiloxane, a silicon-based organic polymer). The conventional lithography is composed of the following process flow: dispensing the material, applying the material on the wafer usually by spinning, and then patterning the definition using a mask aligner and UV illumination. Such technology is suitable for wafer-scale processes both for prototyping (minimum amount is one wafer, possibly comprising several tens or hundreds of microfluidic devices) and for mass-scale production (up to 100,000 wafers of 12″ diameter per year).

Because of the maturity of such technology and existing infrastructure, the method is very popular, and the R&D departments of semiconductor companies and R&D institutions continue to develop new material for this technology (Hegab et al. 2013).

2.11.3 PDMS ENCAPSULATION AND INJECTION

Another very popular technology for microfluidic channel manufacturing is PDMS. It becomes an indispensable material for prototyping and is widely used in a laboratory environment. Indeed, it is easy to process with little or no infrastructure available. The typical process flow is as follows: PDMS precursor (e.g., Sylgard 184 Silicone Elastomer, from Dow Corning; www.dowcorning.com) and a curing agent are mixed at a ratio of 10:1, based on weight. Then, just before the PDMS mixture is casted on the master, the master is coated by an antiadhesive layer for easier removal of the PDMS after curing. The antiadhesive layer can be of a different nature and varies from silicone spray to the layer obtained by silanization (e.g., tridecafluoro-1,1,2,2-tetrahydrooctyl-1-trichlorosilane). Finally, the PDMS mixture is poured onto the master and cured at room temperature (25°C for 48 h) or at an elevated temperature. Typically, it is cured using one of the following curing schedules: at 95°C for 1 h, at 100°C for 35 min, or at 150°C for 10 min. Then, the cured PDMS channel is peeled off from the master and, if necessary, cut according to the specified dimensions and punched to connect the microtubes. The final step is bonding of the PDMS device directly to the substrate (e.g., a glass substrate, PDMS, etc.) with or without any surface treatment.

The abovementioned PDMS precursor Sylgard 184 Silicone Elastomer (Dow Corning) is typically used for electronics and a wide range of industrial applications. In case of the need for biocompatibility, there are biomedical-grade PDMS elastomers available from the same vendor, Dow Corning. For example, SILASTIC MDX4-4210 (Dow Corning) is widely used for medical device encapsulation. It is qualified to address the tests described in ISO 10993-1 for "limited" (less than 24 h) and "prolonged" (less than 3 days) contact duration. The encapsulant meets or exceeds the acceptance criteria. Similar to that, from the same vendor, are the SILASTIC Q7 and C6 series. Similar functional properties have elastomers SILASTIC and Dow Corning LSR series that composed to be applied by injection molding. Complementary to Dow Corning, a vendor, NuSil (www.nusil.com), offers its product portfolio, for example, MED-10xx series, or a two-part material series, MED1-xxxx, MED2-xxxx, and MED3-xxxx product family, with properties similar to those of Dow Corning products.

Despite the abovementioned advantages, because of several reasons (mainly the high cost of the raw material and material functional limitations), generally speaking, PDMS is not suitable for mass-scale production. In addition to that, PDMS is known to cause unwanted contaminations.

2.11.4 PLASTIC MATERIALS

Polymeric materials such as PMMA, PC, PEEK, COC, LPC, POC, PVC, and ABS described earlier are much cheaper and, hence, are said to have limited manufacturability. Because of their relatively low cost, the polymeric microfluidic devices are disposable; the thermoplastic is recyclable and can be remolded and reused many times.

The process flow is conceptually simple and comprises the following steps: processing of microchannel in polymeric materials using a method described below, surface modification before bonding, and finally bonding two parts together. The primary technological procedures used for processing microchannels in polymeric materials are injection molding, hot embossing, micromachining by mechanical micromilling, and photostructuring.

As a mass-scale production technology, macroinjection molding technology, which is advanced and widely used in other applications, can obtain a microchannel of 0.5 mm. For smaller channel dimensions, one must use microinjection molding (with features less than or equal to 0.1 mm), which entails several challenges. The hot embossing technology can be extremely accurate and can result in fine pattern features; hence, in this case, it has a lower throughput and currently is not suitable for mass-scale production. For higher throughput, people use roller embossing technology, which has several limitations: the size of the channel is larger, limited choice of materials, and so on.

Reportedly (Weigl et al. 2001), photostructuring of polymer material is used, which is suitable for prototyping and has limited throughput. Typically CO_2, UV, and IR lasers are used to groove the plastic material, up to a depth of 200 µm. In the case of a deeper microchannel, thin polymer sheets can be processed by the laser technology and there is subsequent lamination to build a 3D structure.

For prototyping and concept proof, mechanical micromilling technology is widely used to obtain channels of 0.2–1 mm in both plastic and soft metals. The technology is cheap and well established.

2.12   LOW-TEMPERATURE ASSEMBLY METHOD FOR A BIOMOLECULE DETECTION SYSTEM

Application of low-temperature technology has resulted in a working prototype of a fully functional biomolecule detection system that performs detection and semi-quantification of influenza A viruses. The fully integrated system is composed of a microfluidic device, an embedded biosensor, and the readout electronics (Loo et al. 2012).

The sensing part consists of interdigitated capacitive electrodes on a $200 \times 200 \ \mu m^2$ area. In order to make the sensor bioactive, biofunctionalization was performed by gas-phase silanization. This process consists of grafting a biological layer on the electrodes. The biological layer is composed of antibodies that have the ability to recognize and bind a specific antigen. In this case, I grafted anti-influenza A antibodies directed against nucleoproteins on the sensing area to capture the influenza A viruses. As the nucleoprotein is essential for the virus survival, this protein is well conserved among different influenza A subtypes. Anti-influenza nucleoprotein antibodies are also labeled with gold beads. If the test sample is infected by the virus, the nucleoprotein will be recognized by the biological layer on the sensor.

Once the labeled antibodies/virus nucleoproteins are attached on the sensor, it causes a capacitance variation that can be measured electrically. The capacitance variation is therefore a function of the number of nucleoproteins captured on the sensor, and it not only indicates the virus presence but also gives a quantitative estimation of the virus concentration.

To build a fully integrated system composed of the microfluidic device, the biosensor, and the readout electronics, I developed an assembly process flow causing no damage to the sensitive biological layer. As described earlier, the use of the standard assembly technique, which entails high temperature, exposure to different chemicals, high pressure, ultrasonic and electromagnetic exposure of a different nature, UV radiation, and so on, causes irreversible damage to the biological layer.

During the first phase of development, the feasibility of biomolecule detection by an interdigitated sensor was demonstrated. The 3×3 mm^2 silicon die has four interdigitated sensing areas and two control ones. The die was assembled on a standard ceramic dual in-line (DIL) package. The assembly process flow consisted of the following steps: die attach, wire bonding, and encapsulation. The die attach was performed with the silver-filled conductive epoxy-based adhesive at room temperature. For the electrical connection between the sensor die and the DIL package, I used Al wire bonding, which is a room-temperature process, unlike the more common Au wire bonding. Finally, the die was encapsulated partially in order to protect the bond pads and the Al wires and to define a fluidic cavity on the sensing area. The intensity of the UV spot is 18.5 mW/cm^2 (maximum irradiance output), the wavelength is 320–500 nm, and the maximum exposition time is 20 s. Such UV exposure did not cause any direct damage to the biofunctionalized layer on the sensor. The fluidic cavity is 3 mm long, 1 mm wide, and 0.5 mm deep, and it allows a volume of test material in the range of 1–1.5 µL (Figure 2.6).

On the second phase of development, a new integrated microfluidic chip with two sensing areas inside microfluidic channels was designed. Two channels that are 300 µm wide are defined on the chip by patterned KMPR. Two inlet reservoirs and one outlet reservoir are seen as 2-mm-diameter holes in KMPR. The dies are fabricated on silicon wafers by photolithography. The interdigitated electrodes are defined in Al and then covered with a very thin layer of alumina (Al$_2$O$_3$). The sensing areas inside the channels are 200×200 µm^2, and the wire bonding pads are 100×100 µm^2, arranged on the lateral sides of the die. The thickness of the Al electrodes and bonding pads is 800 nm. The area around the channels is covered with a 120-µm-thick KMPR photoresist. The silicon dies are 17×14 mm^2. PMMA was selected for the cover material because of its biocompatibility and transparency. The inlet and outlet holes were 2 mm in diameter, micromachined by milling in the 3-mm-thick PMMA cover.

FIGURE 2.6 Biomolecule detection with fluidic channel assembled on the DIL package.

FIGURE 2.7 An assembled microfluidic system.

FIGURE 2.8 The gross-leak test.

FIGURE 2.9 A fully integrated microfluidic system.

The conventional assembly methods for the PMMA cover, such as thermal, plasma activated, or solvent bonding, damage the biological layer on the chip. Therefore, to assemble that, I used a biocompatible, low-temperature (38°C), curable epoxy adhesive (Figure 2.7). To confirm that there is no leak between the die and the cover, I performed a gross-leak detection test using red ink (Figure 2.8). The final assembly step of the biosystem consists of mounting the microfluidic device on a PCB that is part of the whole electronic system (Figure 2.9). The results of the system functionality test were extensively reported (Overstraeten-Schlögel et al. 2011).

2.13 MODELING AND SIMULATION

To finish this chapter, I would like to say a few words of importance in the modeling and simulation of modern microassembly and packaging techniques. The so-called test-and-try-out method is time consuming and is dependent on an engineer's skill and experience. The best way to achieve an optimal solution, reduce the cost and production time, and minimize the impact of the human factor is through modeling and simulation. The use of modeling and simulation will become increasingly necessary for future advances in microassembly and package development. Before the system is designed, the developer must answer the following questions: How will the system be assembled and packaged? What is the system reliability?

Currently, there are a number of modeling studies conducted by research institutions and R&D departments of industrial companies on microassembly and packaging optimization. There are already a number of commercially available software for this.

Currently, there are at least five commercially available multi-physics software (listed alphabetically: ABAQUS, ANSYS, COMSOL, COVENTOR, and OOFELIE) that are reportedly being used for modeling and simulation processes that are part of the microassembly, packaging, and the final system reliability.

Despite the fact that the majority of multi-physics software are continuously being improved, not one of them till now can model and simulate all issues related to microassembly and packaging and affect the final system reliability.

REFERENCES

Blanco Carballo, V. M., Melai, J., Salm, C., and Schmitz, J. 2009. Moisture resistance of SU-8 and KMPR as structural material for integrated gaseous detectors. *Microelectronic Engineering* archive, vol. 86, no. 4–6, April, pp. 765–768.

Chandrasekhar, A., Vandevelde, D. d. V., Driessens, E. et al. 2003. Modeling and characterization of the polymer stud grid array (PSGA) package: Electrical, thermal and thermomechanical qualification. *IEEE Transactions on Electronics Packaging Manufacturing*, vol. 26, no. 1, pp. 54–67, January.

Fjelstad, J. 2007. *Flexible Circuit Technology*, 3rd edition, Seaside, OR: BR Publishing, p. 237.

Fletcher, A. E. 2013. *Advanced Organics for Electronic Substrates and Packages*. Netherlands, Amsterdam: Elsevier.

Gilleo, K. 2005. *MEMS/MOEMS Packaging: Concepts, Designs, Materials, and Process*. New York: McGraw-Hill Nanoscience and Technology Series, p. 239.

Harman, G. 2010. *Wire Bonding in Microelectronics*, 3rd edition. Ohio: McGraw-Hill Professional, p. 446.

Harper, C. A. 2005. *Electronic Packaging and Interconnection Handbook*, 4th edition. Ohio: McGraw-Hill, p. 1000.

Hegab, H. M., Mekawy, A., El, T., and Stakenborg, T. 2013. Review of microfluidic microbioreactor technology for high-throughput submerged microbiological cultivation. *Biomicrofluidics*, vol. 7, no. 2, pp. 021502.

Kakaç, S., Kosoy, B., Li, D., and Pramuanjaroenkij, A. 2010. *Microfluidics Based Microsystems: Fundamentals and Applications*. Germany, Berlin: Springer Science & Business Media, p. 618.

Li, X.-J. J. and Zhou, Y. 2013. *Microfluidic Devices for Biomedical Applications*. Cambridge, UK: Woodhead Publishing (a part of Elsevier), p. 676.

Licari, J. J. and Swanson, D. W. 2011. *Adhesives Technology for Electronic Applications: Materials, Processing, Reliability*. New York: William Andrew Publishing, p. 512.

Liu, S. and Liu, Y. 2011. *Modeling and Simulation for Microelectronic Packaging Assembly: Manufacturing, Reliability and Testing*. Beijing, China: Chemical Industry press, and Singapore: John Wiley & Sons (Asia), p. 288.

Lu, D. and Wong, C. P. 2009. *Materials for Advanced Packaging*. New York: Springer.

Overstraeten-Schlögel, N. Van, Dupuis, P., Magnin, D. et al. 2011. Immunoassay using a biofunctionalized alumina-coated capacitive biosensor: Towards a detection of the H5N1 Influenza virus in microfluidics. *2nd International Conference on Bio-sensing Technology (BITE2011)*, Amsterdam, The Netherlands, October 10–12.

Ratchev, P., Vandevelde, B., and Wolf, I. D. 2004. Reliability and failure analysis of SnAgCu solder interconnections for PSGA packages on Ni/Au surface finish. *IEEE Transactions on Device and Materials Reliability*, vol. 4, no. 1, pp. 5–10, March.

Stoukatch, S., Seronveaux, L., Laurent, P. et al. 2012. Evaluation of aerosol jet printing (AJP) technology for electronic packaging and interconnect technique. *Proceedings of the 4th Electronics System Integration Technology Conferences (ESTC 2012)*, Amsterdam, Netherlands, September 17–20.

Suganuma, K. 2014. *Introduction to Printed Electronics*. New York: Springer, p. 124.

Van Loo, S., Stoukatch, S. Overstraeten-Schlögel, N. Van et al. 2012. Low temperature assembly method of microfluidic bio-molecules detection system. *3rd IEEE International Workshop on Low Temperature Bonding for 3D Integration (LTB-3D 2012)*, Hongo, Japan, May 22–23.

Weigl, B. H., Bardell, R. L., Schulte, T. H., Battrell, C. F., and Hayenga, J. 2001. Design and rapid prototyping of thin-film laminate-based microfluidic devices. *Biomedical Microdevices*, vol. 3, no. 4, pp. 267–274.

Wojciechowski, D., Chan, M., and Martone, F. 2001. Lead-free plastic area array BGAs and polymer stud grid arrays package reliability. *Microelectronics Reliability*, vol. 41, no. 11, pp. 1829–1839.

3 Lab on a Cellphone

Ahmet F. Coskun, Hongying Zhu,
Onur Mudanyali, and Aydogan Ozcan

CONTENTS

3.1 INTRODUCTION

Today, there are approximately 7 billion cellphone subscribers in the world, with a mobile phone penetration rate of ~96% globally [1]. In recent years, there has also been a significant increase in smartphone use especially in the developed parts of the world, which is projected to reach ~40% worldwide by 2015 [1]. Driven by this rapid growth of the mobile phone market, the cost of cellphones has significantly decreased despite dramatic advances in the software and hardware components of these mobile technologies. To this end, the "state-of-art" digital components embedded in cellphones, including image sensors, microprocessors, displays, communication units, and so on, can be employed to create new opportunities for health monitoring in both the developed and the developing regions of the world. Therefore, cellphones, with their built-in features and global connectivity, can provide a ubiquitous platform for biomedical imaging, sensing, and diagnostics applications, which can potentially improve the health care delivery and help reduce the cost of biomedical tests worldwide by enabling the penetration of advanced microanalysis tools to even remote and resource-limited locations.

In this chapter, we review some of our group's recent efforts on developing *lab on a cellphone* platforms (see, e.g., Figure 3.1) that aim to implement multiple telemedicine-related functionalities on cellphones. Some of these emerging cellphone-based technologies that are detailed include (1) transmission microscopes [2,3],

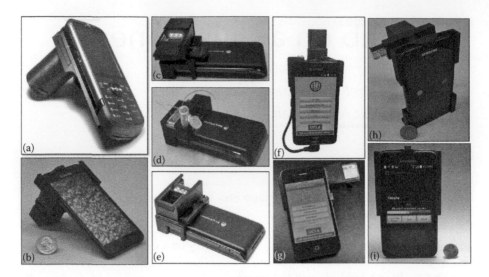

FIGURE 3.1 Mobile phone–based imaging and sensing technologies. (a) Lens-free holographic microscope; (b) multiframe contact microscope, *Contact Scope*; (c) fluorescent microscope; (d) fluorescent imaging cytometer; (e) pathogen (e.g., *Escherichia coli*) biosensor; (f and g) RDT reader; (h) food allergen (peanut) detector; and (i) urinary albumin tester.

(2) fluorescent microscopes and cytometers [4–6], (3) biosensors—particularly pathogen sensing systems [7], and (4) digital readers for rapid diagnostics tests [8–10]. In addition to their local use, we also present the spatiotemporal mapping of health-related information and test results generated through these cellphone-based measurement and imaging technologies toward cloud-based health monitoring/tracking, also providing an important tool for epidemiology [11–13].

3.2 MOBILE IMAGING, SENSING, AND DIAGNOSTIC TECHNOLOGIES TOWARD LAB ON A CELLPHONE

Various biomedical tests are currently performed through the use of benchtop devices, where, for example, bodily fluids or biopsy samples collected from the patients are processed and analyzed using optical microscopes, flow cytometers, plate or microwell readers, among others. These conventional biomedical tools, however, are relatively costly and rather bulky, making them less suitable for use in field or remote settings. In this section, we discuss some of the emerging mobile phone–based imaging and sensing tools that can perform microanalysis on various biological specimens through the use of transmission microscopy, fluorescent imaging and sensing, as well as digital readouts for diagnostics tests.

3.2.1 BRIGHT-FIELD MICROSCOPY ON A CELLPHONE

Optical microscopy has become one of the standard tools in biomedical sciences, especially for disease detection and diagnostics applications. In low-resource settings, for instance, bright-field microscopic evaluation of samples (e.g., sputum, tissue slices, and blood smears) remains as one of the primary methods for the diagnosis of infectious diseases, particularly for tuberculosis (TB) and malaria [14,15]. To create compact, lightweight, and cost-effective microscopic imaging tools, much research has been devoted to the development of mobile phone–based bright-field microscopes [2,3,16,17]. Although the specifications of these transmission microscopes vary based on their designs and working principles, a common feature is that an add-on optome-chanical module is attached to the camera unit of a cellphone for image acquisition, which can then be processed on the same cellphone or transmitted to a central server/ location for further processing, analysis, and reporting. Among these cellphone microscopes, we focus here on two computational methods that can be used to reduce the complexity of these imaging technologies while also enhancing the performance of the optical design: lens-free holographic imaging on a cellphone [2] and multi-frame contact imaging on a fiber-optic array using a smartphone [3].

In the first computational imaging scheme, we demonstrated a lens-free digital microscopy platform on a cellphone as shown in Figure 3.1a [2]. This cellphone-based holographic microscope, weighing ~38 g, does not use any lenses, lasers, or other bulky optical components, enabling a highly compact and lightweight imaging design that can be operated in field conditions. In this platform, we employ partially coherent illumination such as a simple light-emitting diode (LED) that is filtered by a large pinhole (e.g., 0.1–0.2 mm) to illuminate the sample of interest, where the interference of the light waves passing through the micro-objects with the unper-turbed background light creates the lens-free hologram of each micro-object, which is sampled/digitized at the complementary metal-oxide semiconductor (CMOS) detector array that is already embedded in the cellphone camera unit. These raw cell-phone transmission images are then digitally processed through the use of a numeri-cal method (based on, e.g., phase retrieval algorithms) to compute the microscopic images of the samples of interest, yielding both the amplitude and phase information of the samples. Since we operate this imaging system under unit-fringe magnifica-tion, the imaging field of view (FOV) of this lens-free microscopy approach is equal to the active area of the sensor array, typically achieving >20–30 mm² imaging area without the use of any mechanical scanners. We demonstrated the feasibility of this approach by imaging blood cells and waterborne parasites, among other microscopic objects. Such cellphone-based lens-free microscopes, together with their large FOV imaging capability, simplicity, and field portability, hold promise for various tele-medicine applications.

We also created a smartphone-based multiframe contact imaging platform on a fiber-optic array [3], called *Contact Scope*, that can image highly dense or connected samples in transmission mode. This field-portable microscope (see Figure 3.2), weighing 76 g, is also attached to the camera unit of a cellphone through the use of an add-on module, where planar samples of interest are positioned in contact with the top facet of a tapered fiber-optic array. Illuminating the sample using an inco-herent light source such as a simple LED, this transmitted light pattern through the object is sampled by the fiber-optic array. After delivering this transmission image of the sample onto the other side of the taper with ~3× magnification in each direc-tion, this magnified image of the object is projected onto the CMOS image sensor

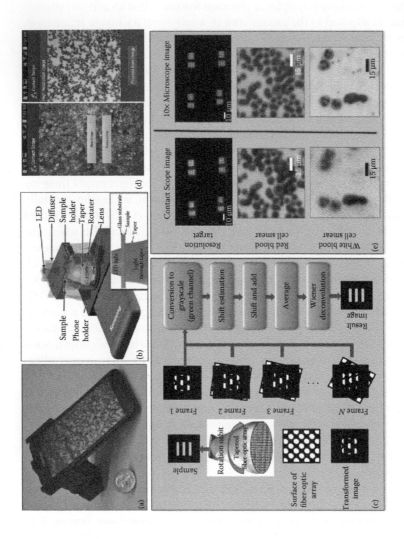

FIGURE 3.2 Smartphone-based multiframe contact microscopy on a fiber-optic array, termed *Contact Scope*. (a) Photograph and (b) schematic of *Contact Scope*. (c) Working principle of multiframe image acquisition and image reconstruction flowchart. (d) Smart Android application running on the same cellphone for digital processing. (e) *Contact Scope*–based imaging of blood smears and corresponding 10x microscope image comparisons. (From Navruz, I. et al. 2013. Smart-phone based computational microscopy using multi-frame contact imaging on a fiber-optic array. *Lab on a Chip* 13(20) (Oct. 21): 4015–4023. Reproduced by permission of The Royal Society of Chemistry.)

of the cellphone through the use of two lenses (one external lens in addition to the cellphone camera lens). Keeping the cellphone camera and the sample at a fixed position, the fiber-optic array is manually rotated with discrete angular increments of, for example, 1° to 2° (see Figure 3.2c). Contact images are captured using the cellphone camera for each angular position of the taper, creating a sequence of transmission images of the same sample. These multiframe images are then digitally combined based on a shift-and-add algorithm through a custom-developed Android application running on the smartphone (see Figure 3.2d), yielding the final microscopic image of the sample, which can also be accessed through the cellphone screen.

We reported the performance of this *Contact Scope* platform by imaging resolution test charts and blood smears as shown in Figure 3.2e, demonstrating that this platform can achieve >1.5–2.5 μm spatial resolution over a FOV of, for example, >1.5–15 mm². This cellphone-based computational contact microscopy platform, being compact and lightweight, also employing rapid digital processing running on smartphones, could be useful for point-of-care applications.

3.2.2 FLUORESCENT MICROSCOPY AND CYTOMETRY ON A CELLPHONE

In addition to the bright-field imaging modalities implemented on cellphones, other optical microscopy techniques including, for example, dark-field microscopy and fluorescence microscopy can also be integrated with cellphones for telemedicine applications. Fluorescent microscopy deserves special attention as it might enhance the sensitivity and specificity in, for example, image-based disease detection and diagnosis. Therefore, combining fluorescent microscopy with cellphone technologies might be rather useful for field-diagnostics needs.

Toward this end, our laboratory has developed a compact and cost-effective cellphone-based fluorescent microscope [5] that can image and analyze fluorescently labeled specimens without the use of any bulky optical components. As shown in Figure 3.1c and d, the major components of this cost-effective platform include a simple lens, a plastic color filter, LEDs, and batteries. In this platform, the excitation light generated by, for example, LEDs illuminates the sample from the side using lens-free butt coupling to the sample cuvette. Because of the large cross section of the LED light, this coupling is quite insensitive to alignment, which makes it repeatable from sample to sample. In this design, the sample holder microchip (where the liquid specimen is located) can be considered to be a multimode slab waveguide, which has a layered refractive index structure. Such a multilayered slab waveguide has strong refractive index contrast at the air–glass (or polydimethylsiloxane [PDMS]) interfaces (i.e., the top and the bottom surfaces), as a result of which the pump/excitation photons are tightly guided within this waveguide. On the other hand, the refractive index contrast at glass (or PDMS)–sample solution interfaces is much weaker compared to air–glass interfaces, which permit a significant portion of the pump photons to leak into the sample solution to efficiently excite, for example, labeled bacteria/pathogens suspended within the liquid sample. Together with this guided wave-based excitation of specimen located within a microchip, the fluorescent emission from the labeled targets is collected by a simple lens system to be imaged onto the cellphone CMOS sensor. For this fluorescent imaging geometry, the image magnification is determined by f/f_2, where f is the focal length of the cellphone camera (e.g., $f \sim 4.65$ mm for the cellphone model Sony-Erickson U10i) and f_2 is the focal length of the add-on lens. Depending on the application and its requirements on resolution and FOV, different magnification factors can be achieved by varying the f_2 value. Note also that this magnification factor is theoretically independent of the distance between the two lenses, which makes alignment of our attachment to the cellphone rather easy and repeatable.

This imaging configuration, when the fluorescent objects are imaged in static mode, is used as a fluorescent microscope as shown in Figure 3.1c. When the fluorescent objects are flowing through the microfluidic channel and their images are recorded in video mode, the device can be used as a fluorescent imaging cytometer [4] to count the fluorescent objects flowing through the microfluidic chamber as illustrated in Figure 3.1d. This scheme can also be used to count cell densities, such as the density of white blood cells in whole blood samples as demonstrated in, for example, the work of Zhu et al. [4].

Quite recently, we have also changed the design of this cellphone-based fluorescent microscopy platform to detect individual virus particles [18]. In this imaging design, a miniature laser diode is used to excite the fluorescently labeled virus particles with an highly oblique illumination angle (e.g., ~75°), which helps us create a decent dark-field background since the low numerical aperture of the cellphone imaging optics misses most of these oblique excitation photons, enhancing our contrast and signal-to-noise ratio so that individual viruses and deeply sub-wavelength fluorescent particles (e.g., 90–100 nm in diameter) can be detected on the cellphone. This single nanoparticle detection and counting capability can especially be important for viral load measurements in field conditions.

3.2.3 Biosensing on a Cellphone

Together with the recent progress in the development of fluorescent imaging modalities on cellphones, we also expanded this cellphone-based fluorescent detection geometry to various biosensing applications, including, for example, the detection of pathogens (e.g., *Escherichia coli*) toward screening of liquid samples such as water and milk [7]). In this scheme (see Figure 3.3), we utilized antibody functionalized glass capillaries as solid substrates to perform quantum dot–based sandwich assay for specific and sensitive detection of *E. coli* in liquid samples of interest (see Figure 3.3b). The *E. coli* particles captured on the inner surfaces of these functionalized capillaries are then excited using inexpensive LEDs, where we image the fluorescent emission from the quantum dots using the cellphone's CMOS sensor utilizing an additional lens that is placed in front of the camera unit of the phone. We then quantify the *E. coli* concentration in liquid samples by integrating the fluorescent signal per capillary tube.

We validated the performance of this cellphone-based *E. coli* detection platform in buffer and fat-free milk samples (see Figure 3.3c and d), achieving a detection limit of ~5–10 cfu/mL in both specimens. Such a pathogen detection platform installed on cellphones, being field portable and cost-effective, could be useful, for example, for rapid screening of water and food samples in low-resource settings.

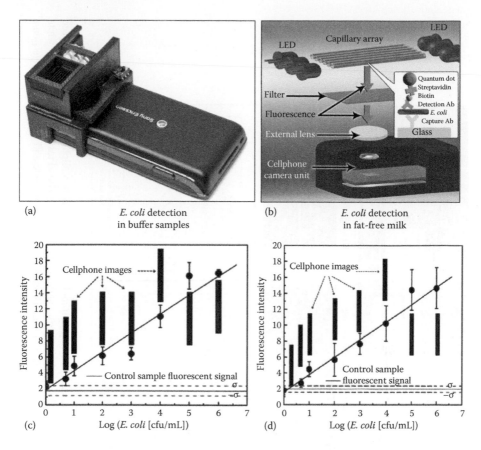

FIGURE 3.3 Detection of *E. coli* on a cellphone. (a) Picture and (b) schematic of the cellphone-based *E. coli* detector. Validation of *E. coli* detection in (c) buffer samples and in (d) fat-free milk specimen. (From Zhu, H. et al. 2012. Quantum dot enabled detection of *Escherichia coli* using a cell-phone. *Analyst* 137(11), (May 7): 2541–2544. Reproduced by permission of The Royal Society of Chemistry.)

3.2.4 BLOOD ANALYSIS ON A CELLPHONE

One of the important applications of cellphone-based imaging and sensing is to perform rapid and cost-effective blood analysis. Toward this need, as shown in Figure 3.4, an automated blood analysis platform running on a smartphone [6] has been created, which consists of two parts: (1) a base attachment to the cellphone that includes batteries and a universal port for three different add-on attachments and (2) optomechanical add-on components for white blood cell and red blood cell (RBC) counting as well as hemoglobin density measurements. Each one of these attachments has a plano-convex lens and an LED source. When a specific attachment is clicked onto the cellphone base, the plano-convex lens in that component gets in contact with the existing lens of the cellphone camera, creating a microscopic

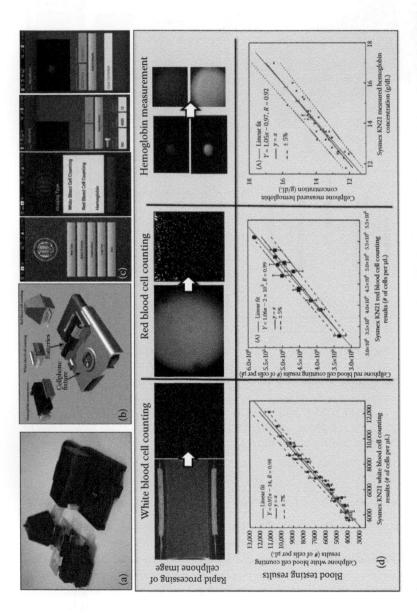

FIGURE 3.4 Rapid and cost-effective blood analysis on a smartphone. (a) Photograph and (b) schematic of the cellphone-based blood analyzer. (c) Android application running on the same smartphone for automated digital processing and quantification. (d) Blood testing results with the white blood cell counts, RBC counts, and hemoglobin concentration measurements. (From Zhu, H. et al. 2013. Cost-effective and rapid blood analysis on a cell-phone. *Lab on a Chip* 13(7) (March 5): 1282–1288. Reproduced by permission of The Royal Society of Chemistry.)

imaging geometry that forms the images of the biological specimen (e.g., blood cells) located within disposable sample holders on the cellphone sensor array. For blood analysis, white blood cells are fluorescently labeled and imaged using an optofluidic illumination scheme as described earlier [5]. Unlabeled intact RBCs are imaged using a bright-field illumination mode, and hemoglobin concentration is estimated based on the measurement of absorbance of the lysed blood sample [6]. A custom-designed Android application is also created to directly process the obtained images on the cellphone (see Figure 3.4). This Android application reports the test results in terms of "number of cells per microliter" and "gram per deciliter" for blood cell counts and hemoglobin concentration measurements, respectively.

We tested the blood samples from anonymous donors to evaluate the performance of this cellphone-based blood analyzer (see Figure 3.4d). For white blood cells, we typically counted 600 to 2500 cells per fluorescent image within a FOV of ~21 mm². The absolute error is within 7% of the standard test results obtained using a commercially available blood analyzer [6]. For RBCs, we counted around 400 to 700 cells per image within a FOV of 1.2 mm². The absolute error for RBC tests is within 5% of the standard test results. For hemoglobin tests, we first measured the transmission intensity of water samples over a FOV of ~9 mm², followed by the measurement of the transmission light intensity of lysed blood samples over the same FOV. We calculated the blood sample absorbance based on these differential measurements. This measured sample absorbance value was then compared to a calibration curve obtained for our platform, and hemoglobin concentration in the unknown sample was estimated based on this linear fitting. This process generated an absolute hemoglobin density measurement error within 5% of the standard test results obtained using a commercially available blood analyzer.

3.2.5 INTEGRATED RAPID-DIAGNOSTIC-TEST READER ON A CELLPHONE

Cellphone-based imaging technologies can also be employed as digital readers for diagnostics tests used especially in developing parts of the world. Since monitoring public health threats is a challenging task in remote locations because of the lack of trained health care personnel and advanced medical instrumentation, the use of rapid diagnostic tests (RDTs), for example, lateral flow-based chromatographic assays, offers significant advantages, superseding clinical examination and other traditional approaches. Special biomarkers in bodily fluids (e.g., saliva, urine, and blood) during an infection or medical condition develop color changes on chromatographic RDTs that are conventionally read via visual inspection by the human eye. However, quantitative and multiplexed chromatographic RDTs have been emerging as part of the next-generation diagnostics tools, enabling highly sensitive and accurate diagnostics beyond the limited functions of qualitative RDTs [19–22].

To provide a compact and cost-effective solution to this important task of quantitative reading of diagnostics tests, cellphones can be utilized to digitize, interpret, store, and transfer the information acquired using these state-of-the-art multiplexed RDTs. Toward this direction, we demonstrated a cellphone-based "universal" RDT reader platform (see Figure 3.5a) that enables digital evaluation of various types of chromatographic RDTs using a lightweight and cost-effective snap-on attachment

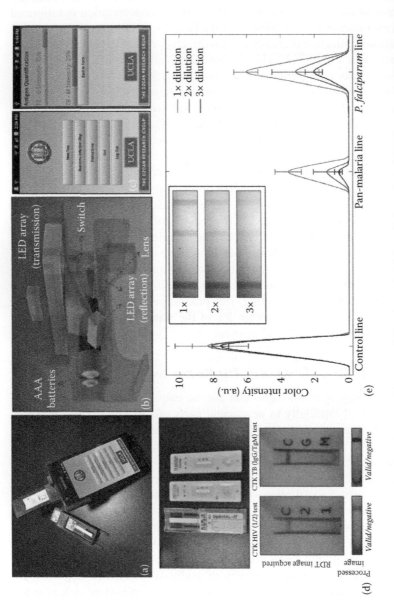

FIGURE 3.5 Integrated RDT reader on a cellphone. (a) Picture and (b) schematic of the cellphone-based RDT reader platform. (c) Custom-developed application running on the same Android phone for digital antigen quantification and validation. (d) Various RDTs (HIV and TB) with their corresponding raw and processed cellphone images. (e) Smartphone-based quantification of malaria RDTs activated using various different concentration levels of antigens. (From Mudanyali, O. et al. 2012. Integrated rapid-diagnostic-test reader platform on a cellphone. *Lab on a Chip* 12(15) (August 7): 2678–2686. Reproduced by permission of The Royal Society of Chemistry.)

(see Figure 3.5b) together with a custom-developed cellphone application (for both iPhone and Android operating systems) as shown in Figure 3.5c [8]. Weighing ~65 g, this cellphone attachment comprises an inexpensive plano-convex lens and multiple diffused LED arrays powered by the cellphone battery through a universal serial bus (USB) cable or two AAA batteries (see Figure 3.5b). This optomechanical unit can be repeatedly attached to and detached from the back of the existing cellphone camera to digitally acquire the images of RDTs in reflection or transmission modes that are processed in real time using the cellphone application to generate a test report including the patient information, test validation, test result, and the quantification of the test signal intensities. The same cellphone application can then store these test reports locally on the cellphone or wirelessly transmit them to a secure database to generate a real-time spatiotemporal map of test results and other conditions (e.g., anthrax attacks or aflatoxin contamination) that can be diagnosed or sensed using RDTs (see Figure 3.6) [8].

Performance of this integrated platform has been demonstrated by imaging various different chromatographic RDTs including Optimal-IT Malaria Tests (Bio-Rad Laboratories, California), HIV 1/2 Ab PLUS Combo Rapid Tests as well as TB IgG/IgM Combo Rapid Tests (CTK Biotech, California) [23–26]. RDTs activated with whole blood samples and positive control wells were successfully imaged using this cellphone-based smart RDT reader to validate the accuracy and repeatability of our measurements (see Figure 3.5d).

In order to demonstrate the sensitivity of our platform, we also imaged Optimal-IT Malaria RDTs that were activated using various different concentration levels of antigens. Starting with the manufacturer-recommended dilution level of Positive

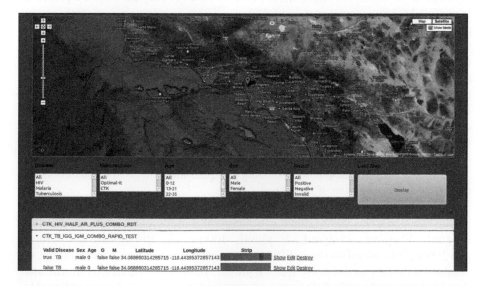

FIGURE 3.6 Spatiotemporal mapping of infectious diseases. Cellphone-based RDT reader–generated test reports and patient data uploaded to a central database. (From Mudanyali, O. et al. 2012. Integrated rapid-diagnostic-test reader platform on a cellphone. *Lab on a Chip* 12(15) (August 7): 2678–2686. Reproduced by permission of The Royal Society of Chemistry.)

Control Well Antigen (PCWA)/20 μL (1× dilution), we diluted the antigens by two, three, and four times to create concentration levels of PCWA/40 μL (2× dilution), PCWA/60 μL (3× dilution), and PCWA/80 μL (4× dilution), respectively, and activated 10 RDTs for each concentration level. Our digital reader imaged and correctly evaluated all the RDTs that were activated with PCWA/20 μL, PCWA/40 μL, and PCWA/60 μL, providing "valid and positive" in the final test reports. On the other hand, because of the weak signal intensity on the RDTs activated with PCWA/80 μL, that is, at 4× dilution compared to the recommended level, the accuracy of our reader dropped to ~60% (see Figure 3.5e).

This cellphone-based RDT reader platform enables digital and quantitative evaluation of various RDTs that are taking the center stage of point-of-care diagnostics in remote locations. Generating a real-time spatiotemporal map of epidemics and other public health threats that can be diagnosed using RDTs, this integrated platform can be useful for policy makers, health care specialists, and epidemiologists to monitor the occurrence and spread of various conditions and might assist us to react in a timely manner.

3.2.6 ALBUMIN TESTING IN URINE USING A CELLPHONE

Some of these emerging mobile phone–based diagnostics platforms can also be utilized for monitoring chronic patients. As an example, chronic kidney disease patients or other patients who suffer from diabetes, hypertension, or cardiovascular diseases might potentially benefit from health screening applications running on cellphones, enabling frequent and routine testing in public settings including, for example, homes, offices, and so on. Along these lines, we have recently developed a personalized digital sensing platform, termed *Albumin Tester*, running on a smartphone [10] that images and automatically analyzes fluorescent assays enclosed within disposable test tubes toward specific and sensitive detection of albumin in urine. Using a mechanical attachment mounted on the camera unit of a cellphone (see Figure 3.7a and b), test and control tubes are excited by a battery-powered laser diode, where the laser beam interacts with the control tube after probing the sample of interest located within the test tube. We capture the images of fluorescent tubes using the cellphone camera through the use of an external lens that is inserted between the sample and the camera lens. These fluorescent images of the sample and control tubes are then digitally processed within 1 s through the use of an Android application (see Figure 3.7c) running on the same phone, providing the quantification of albumin concentration in the urine specimen of interest.

To validate the performance of this cellphone-based *Albumin Tester*, we tested buffer samples spiked with albumin proteins at various concentrations covering 0 μg/mL, 10 μg/mL, 25 μg/mL, 50 μg/mL, 100 μg/mL, 200 μg/mL, 250 μg/mL, and 300 μg/mL. Based on three different measurements for each concentration, we obtained a dose–response curve (see Figure 3.7d) that demonstrates the linear relationship between the spiked albumin concentration in buffer and corresponding relative fluorescent unit (RFU = $I_{test}/I_{control}$, where I_{test} and $I_{control}$ are the *fluorescent* signal of the sample tube and the control tube, respectively) values, achieving a detection limit of ~5–10 μg/mL in buffer. Next, we performed measurements in synthetic urine

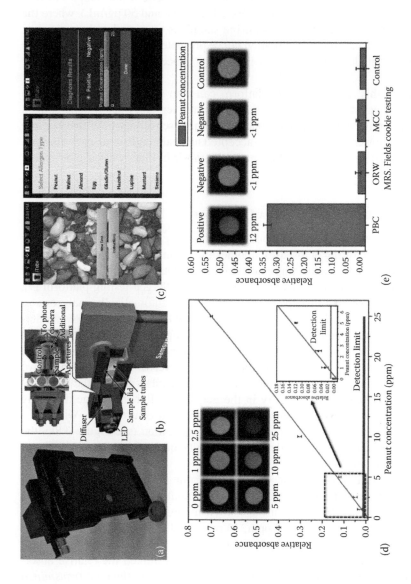

FIGURE 3.7 Allergen testing on a smartphone. (a) Picture and (b) schematic diagram of *iTube*. (c) Custom Android application for allergen quantification on the same smartphone. (d) Calibration of *iTube* based on the experiments performed with samples containing known concentrations of peanut. (e) Testing of commercially available cookies through the use of *iTube*. (From Coskun, A. et al. 2013. A personalized food allergen testing platform on a cellphone. *Lab on a Chip* 13(4) (January 24): 636–640. Reproduced by permission of The Royal Society of Chemistry.)

samples (see Figure 3.7d) spiked with albumin proteins at different concentrations spanning 0 µg/mL, 10 µg/mL, 25 µg/mL, 50 µg/mL, 100 µg/mL, and 200 µg/mL. Similar to our buffer experiments, these experiments achieved an albumin detection limit of <10 µg/mL in urine samples. Furthermore, we also performed blind experiments (see Figure 3.7e) with albumin-spiked synthetic urine samples at randomly selected concentrations (i.e., 200, 150, 25, 10, 100, and 50 µg/mL), where the albumin concentration of each unknown urine sample is estimated with an absolute error of <7 µg/mL based on the calibration curve presented in Figure 3.7d. For these measurements, sample preparation and incubation procedures take approximately 5 min per test.

The users of this *Albumin Tester* platform can also share the test results with their doctors by uploading the data to secure central servers. Such a personalized albumin testing tool running on smartphones, with its sensitivity level that is more than three times lower than the clinically accepted normal range for urinary albumin concentration (~30 µg/mL), could affect early diagnosis of kidney disease or remote monitoring of chronic patients.

3.2.7 FOOD ALLERGEN TESTING ON A CELLPHONE

In parallel to the utilization of cellphone-based imaging and sensing platforms for the diagnosis or monitoring of diseases, another important public concern that could benefit from cellphone-based testing platforms is the detection of allergens in food samples. To this end, we devised a personalized food allergen testing platform [9], termed *iTube*, running on a smartphone that images and automatically analyzes colorimetric assays performed in test tubes for specific and sensitive detection of allergens in food products (see Figure 3.8). Utilizing a compact mechanical attachment installed on the camera unit of a cellphone (see Figure 3.8a and b), we capture the transmission images of our test and control tubes under vertical illumination by two separate LEDs. In this process, allergen assay within the tubes absorbs the illumination light, which changes the intensity in the acquired cellphone image. Digitally processing these transmission images of the sample and control tubes within 1 s through the use of a custom-developed Android application (see Figure 3.8c), the same smartphone application provides the test results, quantifying the allergen contamination in food products.

To demonstrate its proof of concept, we used our *iTube* platform for specific testing of peanut concentration in food products. Initially, we calibrated our *iTube* platform (see Figure 3.8d) by measuring known amounts of peanut concentration spanning 0 ppm, 1 ppm, 2.5 ppm, 5 ppm, 10 ppm, and 25 ppm based on a sample preparation and incubation time of ~20 min per test. Digitally quantifying these samples through the use of the *iTube* platform, we determined the relative absorbance, defined as $A = \log_{10}(I_{control}/I_{test}$, where $I_{control}$ and I_{test} are the total *transmitted* signal through the control tube and the sample tube, respectively), of each test tube, yielding a calibration curve that can be used to quantify the allergen concentration (C) in any given food product of interest based on the measured relative absorbance value (i.e., A) of the target sample.

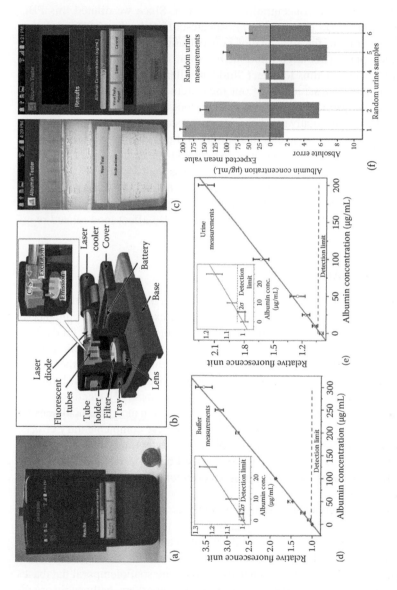

FIGURE 3.8 Urinary albumin testing on a smartphone, termed *Albumin Tester*. (a) Photograph and (b) schematic of the *Albumin Tester*. (c) An Android application installed on the same smartphone for quantification of albumin concentration in the urine sample of interest. Calibration of the *Albumin Tester* based on the measurements performed in (d) buffer samples and in (e) synthetic urine specimen, both of which were spiked with albumin proteins at various concentrations. (f) Urinary albumin testing in randomly selected samples through the *Albumin Tester*. (From Coskun, A. et al. 2013. Albumin testing in urine using a smart-phone. *Lab on a Chip* 13: 4231–4238. Reproduced by permission of The Royal Society of Chemistry.)

Next, we tested three different kinds of commercially available cookies for their peanut concentrations and quantified the level of peanut in these products on the basis of the calibration curve presented in Figure 3.8d. As summarized in Figure 3.8e, our *iTube* platform achieved the following results: (i) Peanut butter chocolate (PBC) cookie was found to be positive (as expected) with an absorbance value of 0.33, equivalent to a peanut concentration of 12 ppm. Since we diluted this PBC extract approximately 5000 times using phosphate-buffered saline (PBS) solution to avoid signal saturation, the real peanut concentration of this PBS sample was more than 60,000 ppm. This large dilution, however, is not required for practical uses as our *iTube* platform aims to detect "hidden" allergen cross-contamination in food products and therefore saturation of our measurement reading is not a practical issue or concern. (ii) Oatmeal raisin with walnut cookie was found to be negative with negligible absorbance, equivalent to a peanut concentration of less than 1 ppm, which implies that walnut existence in this sample did not interfere with our testing results. (iii) Milk chocolate chip cookie was also found to be negative with negligible absorbance, equivalent to a peanut concentration of less than 1 ppm.

Allergic individuals using our *iTube* platform can upload their test results to a central server to create a personalized or public (if desired) testing archive. The spatiotemporal analysis of this allergy database might also affect food related regulations and policies.

3.3   DISCUSSION AND CONCLUSIONS

In this chapter, we presented digital imaging, sensing, and diagnostics platforms running on cellphones that can provide various solutions for existing public health needs and global health problems through microanalysis performed on bodily fluids (e.g., urine, blood), screening of pathogens in liquid samples, or detection of allergens in food samples. In addition to the approaches discussed earlier, several other recent works also provided innovative uses of cellphone-based technologies in biomedical applications, including label-free spectroscopic biosensors [27], electrochemical biosensors [28], pH sensors [29,30], surface plasmon resonance chemical sensors [31], and lab-on-a-card readers [32], among others. All these complementary efforts, aiming to build a multifunctional portable laboratory integrated onto cellphones, might improve access to affordable and personalized health care through state-of-the-art digital imaging, sensing, and diagnostics components and smart applications running on cellphones.

Furthermore, these cellphone-based microanalysis and diagnostics tools, with their connectivity worldwide, can be digitally linked to each other and to central servers, where massive amounts of biomedical data can be securely (although the security and encryption of almost any data can be broken at the cost of computation time) shared and stored in cloud-based networks to create spatiotemporal databases or maps, for example, for various diseases, hazardous biomarkers, pathogenic organisms, and many others. This cloud-based smart global health system might be valuable for telemedicine applications, epidemics, and endemics research, providing a much needed epidemiology tool.

Finally, we should also note that the rapid evolution of cellphone hardware and software and configuration differences from wireless carrier to carrier pose limitations for dissemination and potential commercialization of some of the above-described cellphone-based sensing and imaging technologies. To address this challenge, secondhand or refurbished cellphones can be utilized for the development of these cellphone-based telemedicine platforms as an alternative business strategy, providing device designers and diagnostics companies with a supply of reliable, inexpensive, relatively older-generation phones that can potentially assist the growth and sustainability of the mobile health care market, especially in the developing world, where the smartphone penetration is still at its infancy and the cost of the final product is very sensitive.

Conflict of Interest Statement: A.O. is the cofounder of a startup company (Holomic LLC) that aims to commercialize computational imaging and sensing technologies licensed from UCLA.

ACKNOWLEDGMENTS

The Ozcan Research Group gratefully acknowledges the support of the Presidential Early Career Award for Scientists and Engineers, the Army Research Office (ARO) Life Sciences Division, an ARO Young Investigator Award, a National Science Foundation (NSF) CAREER Award, the NSF CBET Biophotonics Program, an NSF EFRI Award, an Office of Naval Research Young Investigator Award, and the National Institutes of Health (NIH) Director's New Innovator Award DP2OD006427 from the Office of the Director, NIH.

REFERENCES

1. International Telecommunication Union (ITU). 2013. The World in 2013: ICT Facts and Figures. Accessed July 27. Available at http://www.itu.int/en/ITU-D/Statistics /Pages/facts/default.aspx.
2. Tseng, Derek, Onur Mudanyali, Cetin Oztoprak, Serhan O. Isikman, Ikbal Sencan, Oguzhan Yaglidere, and Aydogan Ozcan. 2010. Lensfree microscopy on a cellphone. *Lab on a Chip* 10(14) (June 29): 1787–1792.
3. Navruz, Isa, Ahmet F. Coskun, Justin Wong, Saqib Mohammad, Derek Tseng, Richie Nagi, Stephen Phillips, and Aydogan Ozcan. 2013. Smart-phone based computational microscopy using multi-frame contact imaging on a fiber-optic array. *Lab on a Chip* 13(20) (October 21): 4015–4023.
4. Zhu, Hongying, Sam Mavandadi, Ahmet F. Coskun, Oguzhan Yaglidere, and Aydogan Ozcan. 2011. Optofluidic fluorescent imaging cytometry on a cell phone. *Analytical Chemistry* 83(17) (September 1): 6641–6647.
5. Zhu, Hongying, Oguzhan Yaglidere, Ting-Wei Su, Derek Tseng, and Aydogan Ozcan. 2011. Cost-effective and compact wide-field fluorescent imaging on a cell-phone. *Lab on a Chip* 11(2) (January 21): 315–322.
6. Zhu, Hongying, Ikbal Sencan, Justin Wong, Stoyan Dimitrov, Derek Tseng, Keita Nagashima, and Aydogan Ozcan. 2013. Cost-effective and rapid blood analysis on a cell-phone. *Lab on a Chip* 13(7) (March 5): 1282–1288.

7. Zhu, Hongying, Uzair Sikora, and Aydogan Ozcan. 2012. Quantum dot enabled detection of *Escherichia coli* using a cell-phone. *Analyst* 137(11) (May 7): 2541–2544.
8. Mudanyali, Onur, Stoyan Dimitrov, Uzair Sikora, Swati Padmanabhan, Isa Navruz, and Aydogan Ozcan. 2012. Integrated rapid-diagnostic-test reader platform on a cellphone. *Lab on a Chip* 2012(12): 2678–2686.
9. Coskun, Ahmet F., Justin Wong, Delaram Khodadadi, Richie Nagi, Andrew Tey, and Aydogan Ozcan. 2013. A personalized food allergen testing platform on a cellphone. *Lab on a Chip* 13(4) (January 24): 636–640.
10. Coskun, Ahmet F., Richie Nagi, Kayvon Sadeghi, Stephen Phillips, and Aydogan Ozcan. 2013. Albumin testing in urine using a smart-phone. *Lab on a Chip* 13: 4231–4238.
11. Mavandadi, Sam, Stoyan Dimitrov, Steve Feng, Frank Yu, Uzair Sikora, Oguzhan Yaglidere, Swati Padmanabhan, Karin Nielsen, and Aydogan Ozcan. 2012. Distributed medical image analysis and diagnosis through crowd-sourced Games: A malaria case study. *PLoS ONE* 7(5) (May 11): e37245.
12. Mavandadi, Sam, Stoyan Dimitrov, Steve Feng, Frank Yu, Richard Yu, Uzair Sikora, and Aydogan Ozcan. 2012. Crowd-sourced BioGames: Managing the big data problem for next-generation lab-on-a-chip platforms. *Lab on a Chip* 12(20) (September 18): 4102–4106.
13. Mavandadi, Sam, Steve Feng, Frank Yu, Stoyan Dimitrov, Richard Yu, and Aydogan Ozcan. 2012. BioGames: A platform for crowd-sourced biomedical image analysis and telediagnosis. *Games for Health Journal* 1(5) (October): 373–376.
14. Makler, Michael T., Carol J. Palmer, and Arba L. Ager. 1998. A review of practical techniques for the diagnosis of malaria. *Annals of Tropical Medicine and Parasitology* 92(4) (June): 419–433.
15. Steingart, Karen R., Vivienne Ng, Megan Henry, Philip C. Hopewell, Andrew Ramsay, Jane Cunningham, Richard Urbanczik, Mark D. Perkins, Mohamed Abdel Aziz, and Madhukar Pai. 2006. Sputum processing methods to improve the sensitivity of smear microscopy for tuberculosis: A systematic review. *The Lancet Infectious Diseases* 6(10) (October): 664–674.
16. Breslauer, David N., Robi N. Maamari, Neil A. Switz, Wilbur A. Lam, and Daniel A. Fletcher. 2009. Mobile phone based clinical microscopy for global health applications. *PLoS ONE* 4(7) (July 22): e6320.
17. Smith, Zachary J., Kaiqin Chu, Alyssa R. Espenson, Mehdi Rahimzadeh, Amy Gryshuk, Marco Molinaro, Denis M. Dwyre, Stephen Lane, Dennis Matthews, and Sebastian Wachsmann-Hogiu. 2011. Cell-phone-based platform for biomedical device development and education applications. *PLoS ONE* 6(3) (March 2): e17150.
18. Wei, Qingshan, Hangfei Qi, Wei Luo, Derek Tseng, So Jung Ki, Zhe Wan, Zoltán Göröcs et al. 2013. Fluorescent imaging of single nanoparticles and viruses on a smart phone. *ACS Nano* 7(10) (September 9): 9147–9155.
19. Banoo, Shabir, David Bell, Patrick Bossuyt, Alan Herring, David Mabey, Freddie Poole, Peter G. Smith et al. 2006. Evaluation of diagnostic tests for infectious diseases: General principles. *Nature Reviews. Microbiology* 4(9 Suppl) (September): S21–S31.
20. Vasoo, Shawn, Jane Stevens, and Kamaljit Singh. 2009. Rapid antigen tests for diagnosis of pandemic (swine) influenza A/H1N1. *Clinical Infectious Diseases: An Official Publication of the Infectious Diseases Society of America* 49(7) (October 1): 1090–1093.
21. Mills, Lisa A., Joseph Kagaayi, Gertrude Nakigozi, Ronald M. Galiwango, Joseph Ouma, Joseph P. Shott, Victor Ssempijja et al. 2010. Utility of a point-of-care malaria rapid diagnostic test for excluding malaria as the cause of fever among HIV-positive adults in Rural Rakai, Uganda. *The American Journal of Tropical Medicine and Hygiene* 82(1) (January): 145–147.

22. Wongsrichanalai, Chansuda, Mazie J. Barcus, Sinuon Muth, Awalludin Sutamihardja, and Walther H. Wernsdorfer. 2007. A review of malaria diagnostic tools: Microscopy and rapid diagnostic test (RDT). *The American Journal of Tropical Medicine and Hygiene* 77(6 Suppl) (December): 119–127.

23. Pattanasin, Sarika, Stephan Proux, D. Chompasuk, K. Luwiradaj, P. Jacquier, S. Looareesuwan, and F. Nosten. 2003. Evaluation of a new plasmodium lactate dehydrogenase assay (OptiMAL-IT®) for the detection of malaria. *Transactions of the Royal Society of Tropical Medicine and Hygiene* 97(6) (November 1): 672–674.

24. Moody, Anthony H., and Peter L. Chiodini. 2002. Non-microscopic method for malaria diagnosis using OptiMAL IT, a second-generation dipstick for malaria pLDH antigen detection. *British Journal of Biomedical Science* 59(4): 228–231.

25. Moody, Anthony. 2002. Rapid diagnostic tests for malaria parasites. *Clinical Microbiology Reviews* 15(1) (January): 66–78.

26. Alam, Mohammad Shafiul, Abu Naser Mohon, Shariar Mustafa, Wasif Ali Khan, Nazrul Islam, Mohammad Jahirul Karim, Hamida Khanum, David J. Sullivan, Jr., and Rashidul Haque. 2011. Real-time PCR assay and rapid diagnostic tests for the diagnosis of clinically suspected malaria patients in Bangladesh. *Malaria Journal* 10: 175.

27. Gallegos, Dustin, Kenneth D. Long, Hojeong Yu, Peter P. Clark, Yixiao Lin, Sherine George, Pabitra Nath, and Brian T. Cunningham. 2013. Label-free biodetection using a smartphone. *Lab on a Chip* 13(11) (May 7): 2124–2132.

28. Lillehoj, Peter B., Ming-Chun Huang, Newton Truong, and Chih-Ming Ho. 2013. Rapid electrochemical detection on a mobile phone. *Lab on a Chip* 13(15) (July 2): 2950–2955.

29. Shen, Li, Joshua A. Hagen, and Ian Papautsky. 2012. Point-of-care colorimetric detection with a smartphone. *Lab on a Chip* 12(21) (November 7): 4240–4243.

30. Oncescu, Vlad, Dakota O'Dell, and David Erickson. 2013. Smartphone based health accessory for colorimetric detection of biomarkers in sweat and saliva. *Lab on a Chip* 13 (June 19): 3232–3238. Available at http://pubs.rsc.org/en/content/articlelanding /2013/lc/c3lc50431j.

31. Preechaburana, Pakorn, Marcos Collado Gonzalez, Anke Suska, and Daniel Filippini. 2012. Surface plasmon resonance chemical sensing on cell phones. *Angewandte Chemie International Edition* 51(46): 11585–11588.

32. Ruano-López, Jesus M., Maria Agirregabiria, Garbiñe Olabarria, Dolores Verdoy, Dang D. Bang, Minqiang Bu, Anders Wolff et al. 2009. The SmartBioPhone, a point of care vision under development through two European projects: OPTOLABCARD and LABONFOIL. *Lab on a Chip* 9(11) (June 7): 1495–1499.

4 A Wireless Intraoral Tongue–Computer Interface

Hangue Park and Maysam Ghovanloo

CONTENTS

4.1 INTRODUCTION

4.1.1 WHY TONGUE-OPERATED ASSISTIVE TECHNOLOGY?

Undoubtedly, we are currently living in the era of rapid technological development. Intelligent systems such as smartphone, computer, and robot make our lives more and more efficient, and people can live longer with the development of medical science and improved healthcare technology. However, we should not overlook the fact that individuals with severe disability do not get sufficient benefit from this rapid technological advancement [1,2]. For example, most of the quadriplegic patients still depend on Sip 'n' Puff, head-switch, and mouth-stick, having limited functionality and accessibility, although they are intuitive, robust, and price competitive [3,4]. Moreover, these devices are mostly designed for a single task (e.g., Sip 'n' Puff for driving a power wheelchair (PWC) and mouth-stick for computer access), which necessitates the effort of carrying multiple devices and switching between them.

Researchers have developed several new assistive technologies (ATs) that help disabled people to issue complex commands in a more convenient way. For example, Dragon NaturallySpeaking helps people type long sentences easily with voice recognition and Smartnav helps people control the mouse cursor in any direction and proportionally by tracking the head movement [5–7]. However, they are still limited in computer access and hard to be used for driving PWCs because of the safety issue in noisy environments or the need to locate the tracking device in front of the user's head. Researchers are also developing eye tracking systems such as Tobii or brain signal recognition systems such as Emotiv EPOC to check the possibility of utilizing other remaining abilities of disabled people [8,9]. However, they have limited functionality and also have safety issues (e.g., natural eye movement can be confused with intentional command movement and brain signals can be erroneous without concentration). Accordingly, quadriplegic patients still heavily depend on their family members or caregivers in performing complex tasks or switching devices for different tasks. Clearly, there is a need for ATs that are noninvasive or minimally invasive, easy to use, robust enough to operate well in various environments of daily life, and advanced to easily issue multiple commands and handle multiple devices such as PWCs and computers.

The tongue is well suited for an AT because it is composed of strong and dexterous muscle fibers that do not easily fatigue as long as they are not required to apply force [10]. The tongue and mouth occupy an amount of motor cortex that rivals that of the fingers and the hand [11]. Unlike the eyes, which have rich cortical representations but have been evolved as sensory organs, the mouth and tongue have evolved as motor organs. Thus, they are inherently capable of performing sophisticated motor control and manipulation tasks, which are evident in their role in vocalization and ingestion [11]. Also, the tongue is directly connected to the brain through cranial nerves, which often escape injuries even in severe cases of spinal cord injury [11]. The tongue can even function during random or involuntary neurological activities such as muscular spasms. Therefore, tongue-operated devices are less prone to involuntary movements, which can affect other devices especially those based on electromyography, electroencephalography, or electrooculography signals. Tongue movements are very natural and do not need thinking or concentration. Hence, tongue-operated assistive devices would be easy to learn and use. Furthermore, unlike neural signals from the motor cortex, noninvasive access to the tongue movements is readily available.

Several tongue-operated ATs have been developed to utilize the power of the tongue, such as the tongue-touch keypad (TTK), optical tongue sensing retainer, joystick-based tongue-point, tongue-mouse using pressure sensors, inductive tongue control system (ITCS), and the Tongue Drive System (TDS) [12–17]. Every one of these tongue-operated ATs has its pros and cons in terms of functionality, usability, and acceptability among its potential users. For example, TTK, tongue-point, and tongue-mouse require physical force from the tongue, which may cause fatigue on the tongue, and ITCS and TDS require users to receive a tongue piercing or attach a tracer on their tongue, which may cause temporary discomfort [18]. Nonetheless, researchers are still searching for better solutions to utilize the capabilities of the tongue without sacrificing users' comfort.

4.1.2 WHY WIRELESS INTRAORAL DEVICE?

The intraoral device holds a unique position to efficiently interact with intraoral organs and is recently being developed for several purposes such as blind navigation, tongue stimulation, artificial larynx, and drug diffusion [19–22]. Clearly, the tongue–computer interface is another good application of an intraoral device, because the intraoral device can detect the movement of the tongue at a very short distance. Furthermore, the intraoral device provides two additional advantages for the tongue–computer interface, especially when it is developed as an AT. First, the intraoral device improves the mechanical stability of the tongue–computer interface because the intraoral device can be firmly fixed onto the teeth and sensors maintain their relative position to the tongue. The mechanical stability is important for the AT because it is very important to secure the safety of disabled users. Also, unlike many other ATs that lock the patient in front of a computer monitor or on a wheelchair to be able to use the device, an intraoral device can be used in any position or posture especially if it is wireless. Second, being completely hidden inside the mouth, an intraoral device gives the user a certain degree of privacy and does not become a sign of disability. Disabled people do not want to show their disability to others or to bother their everyday sight with a sign of disability [1,2].

4.1.3 TONGUE DRIVE SYSTEM

TDS is a tongue-operated, minimally invasive, unobtrusive, and wireless AT that infers users' intentions by detecting their voluntary tongue motion and translating them into user-defined commands [19]. TDS can maximally utilize the power of the tongue because of its ability to detect the tongue position in the three-dimensional (3D) intraoral space, which allows users to define a virtually unlimited number of commands by moving their tongues to various desired locations, as long as they can remember and repeat those tongue gestures. In addition, the TDS does not need any steering force or pressure applied by the tongue, which can avoid fatigue of the tongue unlike several other tongue-operated ATs using the physical strength of the tongue [13,15,16,18].

The TDS was first developed as a headset, referred to as the external TDS (eTDS), with a pair of poles locating the magnetic sensors near the cheeks to measure the tracer magnetic field with sufficient signal-to-noise ratio (SNR) [23]. The eTDS headset has been evaluated, and its performance has been compared with a popular AT, called Sip 'n' Puff, through two rounds of clinical trials in two rehabilitation centers [24–27]. To improve the mechanical stability of the system and better protect the users' privacy, the intraoral version of the TDS (iTDS) has been designed to locate inside the mouth on a dental retainer that clasps onto the teeth [12,28,29]. Accordingly, the iTDS works as an efficient tongue–computer interface by taking advantage of both the power of the tongue and the strength of the intraoral device. In Sections 4.2 through 4.5, this book chapter describes a design procedure and techniques for a wireless intraoral tongue–computer interface, by using the iTDS as a practical example.

4.2 SENSOR/ANALOG/DIGITAL

4.2.1 APPLICABLE SENSOR TECHNOLOGIES FOR THE TONGUE–COMPUTER INTERFACE

Sensor technology is necessary to detect tongue movement. Various kinds of sensors have been recently developed in a small packaging by employing microelectromechanical systems technology, and system designers should choose one of them according to the system budget and target application. The electromechanical switch, which converts the physical force from the tongue to the necessary electrical signal, would be one of the most intuitive sensor technologies. Steering switch or push button switch has been applied to joystick-based tongue-point or TTK [13,15]. Although such a switch-based intraoral tongue–computer interface provides easy user interface and high robustness, it suffers from the limit in the number of command or causes fatigue on the tongue. Considering that intraoral space is a light-sensitive environment because of the darkness inside the mouth and the tongue is thick enough to distort the path of light, an optical sensor can be an option to detect the tongue movement [14]. However, for the intraoral devices employing optical sensors, it would not be easy to avoid the "Midas touch problem" because the tongue occupies almost 80%–85% of the oral cavity [30]. System designers on the optical sensor–based intraoral tongue–computer interface must be very careful for the spatial sensor arrangement and the sensitivity of each sensor.

4.2.2 MAGNETIC SENSOR–BASED TONGUE–COMPUTER INTERFACE

Magnetic sensors, such as the ITCS and the iTDS, have been recently employed in the intraoral tongue–computer interface [12,17]. Using DC magnetic field for either wireless manipulation or displacement sensing is desired in medical devices because no power is necessary for generating magnetic fields from permanent magnets. Yet another advantage of the DC and low-frequency magnetic fields is that the human body is made of nonmagnetic material and completely transparent to such magnetic fields. Furthermore, the magnetic field–based detection allows the tongue to move freely without any burden of applying steering force or pressure and allows users to define virtually unlimited number of commands by utilizing the entire intraoral space, as described in the explanation of the TDS. However, a magnetic tracer needs to be attached to the tongue to generate a magnetic field by adhesive/piercing/implantation, which might act as a hurdle for the magnetic sensor–based intraoral tongue–computer interface to be selected by the users.

There are several types of magnetic sensors such as the Hall-effect sensor, anisotropic magnetoresistance (AMR) sensor, tunnel magnetoresistance sensor, giant magnetoresistance sensor, and flux-gate sensor, most of which have been implemented in small and low-power modules that include interfacing and compensation circuitry for positioning and mobile navigation applications [31]. Each type of magnetic sensor has pros and cons in terms of size, sensitivity, and price, and system designers need to select the appropriate magnetic sensor according to their system budget and target application. For the iTDS, we adopted a three-axial AMR-type magnetic sensor, HMC1043 (Honeywell, Morristown, New Jersey), because of its small footprint and low power consumption [32]. AMR sensors are arguably one of the top candidates because of their sensitivity range and sampling bandwidth, which are sufficient for tracking a small permanent magnetic tracer, attached to the tongue, within the 3D oral space. The HMC1043 contains three AMR Wheatstone bridges (one per orthogonal axis) for achieving high linearity, plus internal nulling mechanisms to cancel offset and drift. Four 3-axial HMC1043 sensors were placed on each corner of the printed circuit board (PCB) to provide sufficient coverage of the intraoral magnetic field variations. It can be shown that these 12 sensor outputs can provide sufficient information for detecting the position and the orientation of a magnetic dipole tongue after cancelling the earth's magnetic field interference [33].

Note that the magnetic sensor technology is still in a very active development cycle and major magnetic sensor companies such as Honeywell, Freescale, and STMicroelectronics recently offer low-power digital 3D magnetic sensors in ultrasmall packages. Those companies either produce analog magnetic sensors no more or do not make any revision for them, which results in a high price and a large packaging compared to the digital magnetic sensors. Although the analog magnetic sensor such as HMC1043 provides more design flexibility with an analog output, the digital magnetic sensor would be the right choice for the system design in the future.

4.2.3 Analog Front-End

Differential outputs of the HMC1043 bridges need to be amplified via precision instrumentation amplifiers to levels that would be sufficient for analog-to-digital conversion, which has a full-scale input dynamic range of ground to V_{DD}. Considering the HMC1043 sensitivity of 1 mV/V/gauss and the strength of the field generated by the magnetic tracer at sensor locations (0.1–10 gauss), four differential gain settings of 25, 50, 100, and 200 V/V were selected.

The sampling rate of the analog front-end (AFE) was determined considering the subject's reaction time, tongue dwelling time at each command position, and oversampling ratio of the magnetic sensor data. From our previous trials with the eTDS, the subjects' reaction time was 0.2–0.5 s; that is, subjects maintained their tongues at command positions for 0.2–0.5 s before moving to another command position [34]. The magnetic sensor data need to be oversampled at each command position to filter out the unintended tongue vibrations. We chose a 50-Hz sampling rate for the eTDS to take at least 10 samples at each command position, and its efficacy was empirically verified in human subject trials [25,35,36]. In the case of the iTDS, a 64-Hz sampling rate was selected as the closest number to 50 Hz by dividing the 32.768-kHz iTDS system clock frequency by a factor of 512. The detailed sampling timing of the AFE and analog-to-digital converter (ADC) is described in Figure 4.1a. Circuit level description of the AFE has been summarized in the following paragraph, but further details can be found in Ref. [37].

The 12×2-to-3×2 fully differential time division multiplexing (TDM) multiplexer selects the sensor to be processed and feeds the differential outputs of $x/y/z$ axes to three AFE channels. In each AFE channel, the sensor bridge output is amplified by a fully differential current-feedback instrumentation amplifier (CFIA), providing an electronically selectable gain of 5, 10, 20, or 40 V/V, while achieving offset cancellation. The CFIA, shown in Figure 4.1b, employs auto-zeroing to provide high input impedance, low noise, high linearity, high common-mode rejection ratio, low offset/drift, and precise and stable gain by means of isolation and balancing [38]. The circuit mainly consists of an input transconductance stage (G_{m1}) and a differential output transimpedance stage (R_m). A voltage follower (M_5, M_6) preceding G_{m1} accommodates with the common-mode voltage at the magnetoresistive sensors' output (V_{ip}, V_{in}), which is $V_{DD}/2$, while driving the gates of the 0.5-μm pMOS input devices (M_1, M_2) at a suitable DC level. The current through the input diff-pair (M_1, M_2) is held constant by a feedback loop consisting of G_{m2} and current sources M_7–M_{10}. Feedback through G_{m2} balances G_{m1} and forces the differential input voltage to fall predominantly across R_1, resulting in low distortion. By mirroring the balancing currents onto the transimpedance stage, the circuit achieves high isolation, and the gain can be precisely set by the resistor ratio, R_2/R_1, which is stable across process corners and temperatures. Moreover, opamp A_1 provides common-mode feedback for the transimpedance stage.

Input devices have a large W/L ratio to achieve high transconductance with limited thermal noise, whereas $1/f$ noise has been reduced by employing devices with large $W \cdot L$ wherever possible. Offset cancellation was achieved through an auto-zeroing scheme, which consists of sampling the output of the CFIA between each readout, after shorting the inputs by nMOS switches. Then, a correction current was applied in the transimpedance stage through feedback at the auxiliary port V_{os}, which drives the gate of M_{20} and maintains it until the next readout to balance this stage.

The amplified signal outputs from the AFE are then low-pass filtered by a linearized pseudo-RC filter, limiting broadband noise and aliasing, and then further amplified by a differential-to-single-ended amplifier presenting a fixed gain of 5 V/V, before being sampled at 64 Hz through an S/H amplifier featuring clock feedthrough cancellation. Finally, the output samples from the three readout channels are multiplexed one more time by another 3-to-1 TDM multiplexer and buffered before digitization in the following ADC.

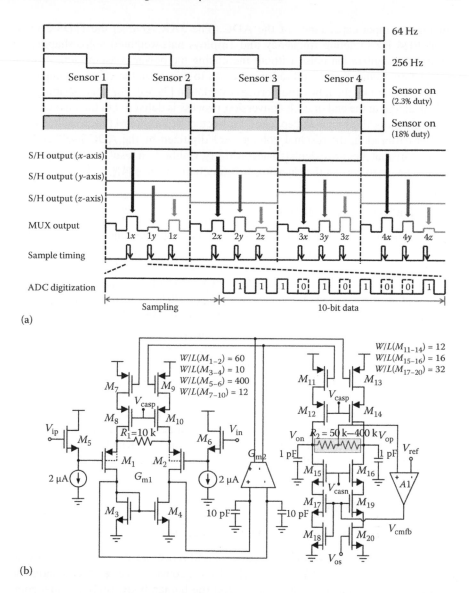

FIGURE 4.1 (a) Timing diagram of the AFE and ADC with adjustable duty cycling and (b) schematic diagram of the CFIA with offset cancellation, used at the input of the iTDS AFE.

4.2.4 Analog-to-Digital Converter

The iTDS can either use the off-chip ADC or on-chip ADC according to its power budget. The first version of the iTDS (iTDS-1) used the ADC built in the microcontroller (MCU) and fed the amplified analog output to the MCU to get the digitized data with an appropriate header for the wireless transmission. The second version of the iTDS (iTDS-2) has an on-chip 13-bit successive approximation register (SAR) ADC to reduce power consumption and loading effect on the AFE output stage by reducing the input capacitance of the ADC. The SAR ADC at the iTDS-2 operates at 1024 Hz sampling frequency and employs half-weighted noise shaping, to decrease the digitization noise without increasing the power consumption. Also, it employs foreground digital calibration schemes to minimize the effects of capacitor mismatch caused by the fabrication process variation [39,40]. For noise shaping, the residual charge at the end of each conversion process is delivered to the next conversion stage, using the switched capacitor connected in parallel to the capacitor bank [39]. For calibration, the residual voltages are digitized by the ADC itself and the capacitor mismatches are calculated accordingly. Calibration results are stored in the microprocessor to correct the capacitor mismatch in advance [41].

4.2.5 DIGITAL CONTROL BLOCK

To reduce the size and complexity of the digital control block and increase the flexibility of control signals, an ultralow-power MCU or field-programmable gate array (FPGA) can be used both as off-chip components. In the actual implementation of the iTDS-1 and the iTDS-2, we used MSP430 MCU (Texas Instruments, Dallas, Texas) and AGLN250 FPGA (Microsemi, Aliso Viejo, California), respectively. Since MSP430 already has an embedded ADC, it eliminates the need for such a block in the iTDS system-on-a-chip (SoC). AGLN250 must be also considered to reduce the power consumption, because it still has the ability for digital signal generation and a customized ADC consumes much less power than the ADC inside MSP430. Note that AGLN250 can handle faster data rate than MSP430, which can be beneficial for the digitally assisted super-regenerative receiver introduced in Section 4.3 [42]. For assisting wireless transmission, either MSP430 or AGLN250 can generate an appropriate data packet including a header with a predefined data rate up to ~1 Mbps, which is sufficient for radio-frequency (RF) transmitters working at 433.9 MHz located in the industrial, scientific, and medical (ISM) radio band.

4.3 WIRELESS LINK

4.3.1 WIRELESS COMMUNICATION METHOD SELECTION

The explosive growth of wearable electronics and recent advancements in implantable medical devices (IMDs) have coincided with considerable research conducted on wireless communication inside and around the human body. Body channel communication (BCC) has been proposed as a power-efficient wireless communication method for devices that are in contact with the human body [43]. Two common BCC scenarios, capacitive and galvanic BCC, have been tested by researchers and verified to work in 30–70 MHz and 0.01–1 MHz bands, respectively [43–45]. Utilizing the conductivity of the human body, capacitive BCC works according to the principle of quasi-static near-field coupling. That is, it must have a return path for the current, which is usually formed through capacitive coupling with common ground, for example, the earth [43]. Galvanic BCC uses an alternating current initiated from a pair of differential transmitter (Tx) electrodes, distributed in the surrounding tissue, and picked up by the receiver (Rx) electrodes [45,46]. Because this method does not need ground coupling, it can be used for communication inside the body. However, galvanic BCC is suitable for implant-to-implant communication, in which both Tx and Rx electrodes are inside the body. It shows high attenuation when the signal meets the skin because the skin is highly resistive to alternating current [45,46] and is completely ineffective when the Rx is detached from the body.

As intraoral devices are in contact with the gums, the tongue, or the palate, and their associated receivers may be in contact with the human body, BCC can be considered to be used for these devices. However, when applied to intraoral devices, neither of the BCC methods maintains their desired operating conditions. For capacitive BCC, coupling between the Tx and the common (earth) ground is severely attenuated because the Tx has to be located inside the mouth, as in the case of IMDs [46]. When galvanic BCC is applied to the intraoral device, high attenuation at a skin is expected because the Tx and the Rx is located on or outside the body. Moreover, both types of BCC require special electrodes to minimize attenuation at the boundary between the device and the body.

Unlike BCC, RF, which is widely used in wireless communication, does not require special operating conditions. It has been successfully adopted in IMDs such as the pacemaker and neurostimulator, and considerable ongoing research is modeling and characterizing the electromagnetic (EM) aspects of IMDs [47–56]. Because intraoral devices are located inside the body, they have EM characteristics similar to those of IMDs. However, designing a wireless link for an intraoral device could be more challenging than it is for the IMD link because, depending on its anatomical location, the latter is often surrounded by a distinct and fairly stable tissue environment. By contrast, the intraoral device is located in a constantly changing environment depending on the relative positions of the jaws and movements of the tongue, which continuously changes shape when one swallows, breathes, or speaks. The lower jaw position also changes during speech and ingestion.

4.3.2 Frequency Band Selection

To employ RF communication for intraoral devices, we must choose a carrier frequency because of performance degradation from three main sources that are highly frequency dependent. One source is the size of an antenna, which is limited so that it comfortably fits inside the mouth and does not interfere with natural functions such as breathing, ingestion, and speech. The size reduction of the antenna lowers its radiation or coupling efficiency, particularly at lower frequencies with longer wavelengths. Another source is the human body, including the gums, the teeth, and the lips, which surrounds the electronics and attenuates the RF signal, more at higher frequencies. The third source is the impedance mismatch between the power amplifier (PA) and the antenna. Change of geometry, thickness, and proximity of the tissue around the electronics can distort the impedance matching, particularly at higher frequencies [57,58].

To determine the optimal carrier frequency for the intraoral device, we selected three candidates: 433.9 MHz, 2.48 GHz, and 27 MHz. All three bands are located in the ISM band and available for unrestricted usage. Among the ISM frequencies that behave as propagation waves at distances below 1 m, 433.9 MHz is the lowest with relatively small attenuation in the human body. This frequency is also close to the medical implant communication service 402–405 MHz band; therefore, it is expected to show similar characteristics. We chose 2.48 GHz because of its good antenna radiation efficiency in small-sized antennas, considering that its quarter wavelength is ~3 cm. We also chose 27 MHz to employ a near-field communication by inductive coupling, which can be more efficient than EM wave propagation at short distances. In addition, the human body is almost transparent to the near-field communication at 27 MHz because its wavelength is >10 m and it mainly depends on the magnetic field [59].

Measurement results from path loss experiments, antenna radiation patterns, and coil coupling patterns show that the 433.9-MHz band is the best choice among the three selected frequencies for the iTDS communication from inside the mouth [60]. Even though the 27-MHz band shows the best performance for distances <39 cm because of its negligible loss in the tissue and strong inductive coupling, it is highly sensitive to distance (slope of the curve in Figure 4.2a) and angular alignment between the Tx and the Rx coils because of their near-field interactions. However, we

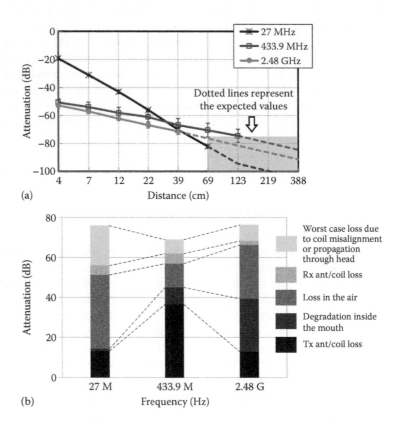

FIGURE 4.2 (a) Measured path loss between Tx antenna/coil transmitting from inside the mouth (closed) and Rx antenna/coil in the air, at 27-MHz, 433.9-MHz, and 2.48-GHz bands and (b) composition of power loss at 27-MHz, 433.9-MHz, and 2.48-GHz bands, measured at 22 cm from the iTDS inside the mouth.

still keep the 27-MHz band in the iTDS along with the 433.9-MHz band to improve the system robustness by employing the dual-band radio. Also, we expect to find a good location around the head (e.g., head rest of the desk chair or PWC) to locate a 27-MHz receiving coil, to secure a robust wireless communication at the 27-MHz band in a short distance. The 2.48-GHz band, showing the best performance when the iTDS is operating in the air, is severely degraded when the iTDS is worn inside the mouth and results in inferior performance compared to the 433.9-MHz band. Figure 4.2b depicts the composition of the loss sources for each band at 22 cm Tx–Rx separation. This figure shows that the major source of power loss varies in each band.

At marginal operational distances obtained from a bit error rate (BER) test with 0 dBm output power, we conducted a link budget analysis to check if the result of the BER test matches with the result of path loss experiments and to investigate the composition of power loss at each band. For each band, we measured BER at nine logarithmic distances used in the path loss experiments and determined a marginal operational distance as the maximum distance satisfying BER $< 10^{-7}$ [60]. Table 4.1 shows the result of link budget analysis for three candidate carrier frequencies, based on the values of each source of power loss obtained by either experimental results or theoretical calculations.

TABLE 4.1

Link Budget Analysis of the iTDS at 27 MHz, 433.9 MHz, and 2.48 GHz

Carrier Frequency	27 MHz	433.9 MHz	2.48 GHz
Data rate (kbps)	57.6	250	250
Modulation	OOK	GFSK	MSK
Max. Tx–Rx distance (cm)	22	123	39
Output power (dBm)	0	0	0
$L_{Feed/Balun}$ (dB)	−4.00	−2.89	−3.32
$L_{Tx.IMN}$ (dB)	−9.50	−3.11	−5.68
$L_{Tx.ANT(Coil)}$ (dB)		−26.85	−1.82
$\Delta L_{Tx.IMN} + \Delta L_{Tx.ANT(Coil)} + L_{HB}$ (dB)	−1.00	−9.72	−26.64
L_{AIR} (dB)	−37.00	−26.98[a]	−31.98[a]
$L_{Rx.ANT(Coil)}$ (dB)	−4.50	−4.95	−1.96
Received power at Rx ANT/coil (dBm)	−56.00	−74.50	−71.40
$L_{Feed/Balun}$ (dB)	−4.00	−2.89	−3.32
Worst-case L_{Angle} (dB)	−18.00	−5.84	−8.20
Fading margin (dB)	12.50	11.77	7.08
Rx sensitivity (dBm)[b]	−90.5	−95.0	−90.0
Required SNR for BER $< 10^{-7}$ (dB)[b]	12.0	–	–
Rx noise figure (dB)[b]	17.5	–	–
Noise power (dBm)	−120	−116.5	−116.5
Channel filter BW (kHz)	250	540	540
Thermal noise (dBm/Hz)	−174	−174	−174

[a] Calculated from the Friis' transmission formula [57].
[b] Calculated/derived from datasheets [60].

4.3.3 Dual-Band Transceiver Design

To improve the robustness of the wireless link in the face of strong interference, the iTDS incorporates a dual-band radio, and the iTDS-2 even has a functionality to automatically switch between 27 MHz and 432 MHz without user input [61]. The wireless link quality is detected by the built-in super-regenerative receiver (SR-Rx) in the iTDS-2 and a CC1110 transceiver (Texas Instruments) in the TDS universal interface (TDS-UI; see Section 4.5.4), for 27 MHz and 432 MHz, respectively. The SR-Rx detects the interference at 27 MHz when the strength of the interference is higher than −73 dBm, the sensitivity of the SR-Rx. The CC1110 detects interference at 432 MHz when the packet error rate is above 0.3%, which corresponds to the BER of 10^{-5} for 296 bits in a packet. When CC1110 detects the interference, the 27-MHz Tx in the TDS-UI sends an 8-bit Ack packet to the iTDS-2 for three times. When the SR-Rx in the iTDS-2 detects the 27-MHz carrier, the FPGA stores the incoming 8-bit data stream, continuously updates the data in a first-in first-out buffer, and compares it with a predefined Ack sequence (10100110) to determine whether the input is an Ack packet or a mere interference. If the FPGA recognizes an Ack packet, it changes the communication frequency from 432 MHz to 27 MHz, and vice versa, if an Ack packet is not detected in the presence of 27 MHz interference.

Figure 4.3a and b describe how the dual-band data link between the iTDS-2 and the TDS-UI works, as a block diagram and a timing diagram, respectively. For

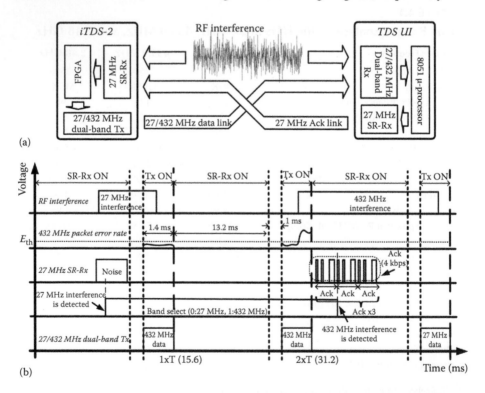

FIGURE 4.3 Dual-band data link operation at 27/432 MHz and the Ack link at 27 MHz: (a) block diagram and (b) timing diagram.

13.2 ms time interval between data transmissions, the SR-Rx monitors interference at 27 MHz and receives information about interference at 432 MHz via the 27-MHz Ack link. The recovered data packet by the SR-Rx is fed into the FPGA to switch the communication band between 27 MHz and 432 MHz. The left half of Figure 4.3b shows a transition from 27 MHz to 432 MHz and the right half shows a transition from 432 MHz to 27 MHz. The dual-band Tx transmits the magnetic sensor data within 1.4 ms via the selected communication band. Note that we assume at least one of the communication bands has a good wireless link quality. The dual-band Tx will continuously switch the band if both communication bands have poor wireless link quality below the threshold.

Figure 4.4a shows a simplified schematic diagram of the dual-band low-power Tx. Two crystal oscillators, operating at 27 and 48 MHz, have been utilized to generate an accurate timebase, using off-chip crystals. The 432-MHz carrier is generated by multiplying the 48-MHz clock by 9, utilizing a dual-loop delay locked loop (DLL) and an edge combiner [62]. The 27-MHz Tx includes a class C PA preceded by a 27-MHz crystal oscillator, as shown in Figure 4.4a.

The DLL-based topology for the 432-MHz Tx is attractive because of its lower power consumption, better stability, less jitter, and faster locking time compared to phase locked loops, in which a power-hungry voltage-controlled oscillator is needed to create the high-frequency carrier. The edge combiner behaves like a nonlinear PA and injects sufficient current to the off-chip LC tank, which matches the 50-Ω antenna to the Tx output. The dual-loop DLL includes nine voltage-controlled delay cells, which have been designed with current-starved inverters, phase-frequency detectors (PFDs), and charge pumps. Two identical feedback loops delay the rising and falling edges of the 48-MHz clock in each cell to generate nine 50% duty cycle, equally spaced clocks.

A schematic of the dynamic PFD, which is used to control the charge pump, is shown in Figure 4.4b [12]. The PFD can be divided into two halves that are identical except for input signals that are switched. Each block consists of two cascaded stages with a precharge pMOS in each stage. Dynamic PFD eliminates flip-flops in conventional PFDs, which dissipate high power at higher clock frequencies, while maintaining a simple structure and fast transition times. The specific charge pump topology that is shown in Figure 4.4b was chosen to reduce charge injection, clock feed-through, and charge sharing in conventional charge pumps. The source and sink currents are carefully matched by increasing the length of current mirror transistors, M_1 and M_2, and adding a replica branch (T_3 and T_4 switches) of the charge/discharge branch (T_1 and T_2 switches) to avoid switching of the current source transistors.

4.3.4 Adaptive Impedance Matching

To minimize the signal reflection between the transmitter and the antenna in the time-varying environment such as an intraoral space, we tested an adaptive impedance matching technique at 433.9 MHz. We do not include 27 MHz here because the human body, including varying intraoral environments, has a minimal effect on near-field inductive coupling at 27 MHz [63]. To minimize reflection at the interconnect, impedances toward the balun (Z_1) and the antenna (Z_2), as shown in Figure 4.2a,

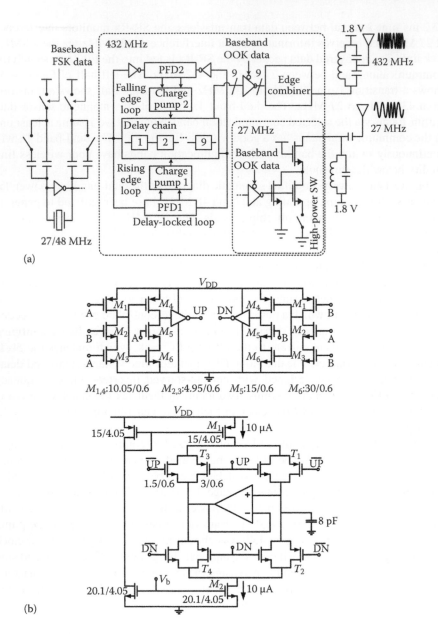

FIGURE 4.4 (a) Schematic diagram of the dual-band Tx that operates at 27 or 432 MHz and transmits data by applying either OOK or FSK and (b) schematic diagram of the dynamic PFD and charge pump.

need to be conjugate matched with each other [57]. To verify the adaptive matching scheme for 433.9 MHz, we used a commercial chip antenna ANT1204F002R0433 (Yageo, San Jose, California) having a small footprint to be mounted on the iTDS board. We also chose 0433BM15A0001 (Johanson, Camarillo, California) as a balun to connect differential output to single antenna feedpoint. In the air, Z_1 and Z_2 are well matched because the selected balun has a 50-Ω unbalanced port impedance and the selected antenna is also matched to 50 Ω at the target frequency.

However, in an intraoral space, Z_1 and Z_2 become mismatched because Z_2 changes considerably in dynamic intraoral environments in the case of small-sized chip antennas while Z_1 remains almost the same. To address this problem, we employed an adaptive matching network that dynamically adjusts the Z_2 in response to changes of the intraoral environment [64,65]. The adaptive matching network consists of a switched CLC pi matching network and a power detector. A power detector generates voltage according to the power at the antenna port and the MCU finds the optimum setting of the CLC pi matching network by sweeping the switch settings and monitoring the power detector output. To deliver a small portion of the power at the antenna port to the power detector with minimal loading effect, a directional coupler is connected between the matching network and the antenna, as in Figure 4.5a. We chose MDC201 (MACOM, Lowell, Massachusetts) as directional couplers.

The CLC pi-network components, shown in Figure 4.5b, are carefully selected to cover the matching range for the varying environments inside the mouth. Note that we used a fixed inductor value by compromising each CLC combination because the switch at the inductor could generate an undesirable parasitic effect and totally change the CLC network characteristics. While the MCU controls four field effect transistor (FET) switches in all 16 possible combinations, a power detector feeds the voltage output into the built-in ADC of the MCU. The power detector connected to a coupling port of the directional coupler monitors output power with minimal effect onto the original output. The MCU digitizes the power detector output for each of the 16 switch settings, finds out the one that generates the highest value, and fixes four switches at that configuration. Figure 4.5c shows the exemplary output voltage of the power detector during the adaptive matching. On the left side of the waveforms, the Tx sends data with unmatched setting when the matching condition is distorted by the change of intraoral environment. During the optimization period shown in the middle of the waveforms, the adaptive matching monitors the power detector output with 16 switch settings and updates an optimal switch setting, which maximizes the output power at the antenna port, as shown on the right side of the waveform. The optimization period takes ~30 ms and is repeated in every 1 s to update the optimum configuration according to changes in the intraoral environment. Because the iTDS sends data in every 20 ms and cannot send data normally during the optimization period, the adaptive matching may degrade the real-time operation of the iTDS [66].

4.3.5 ANTENNA DESIGN

For the 432-MHz band, data are delivered by EM propagation at a nominal range of 0.3–2 m for environmental access and wheelchair navigation [12,67]. The antenna geometry should be carefully designed to optimize the wireless link propagating

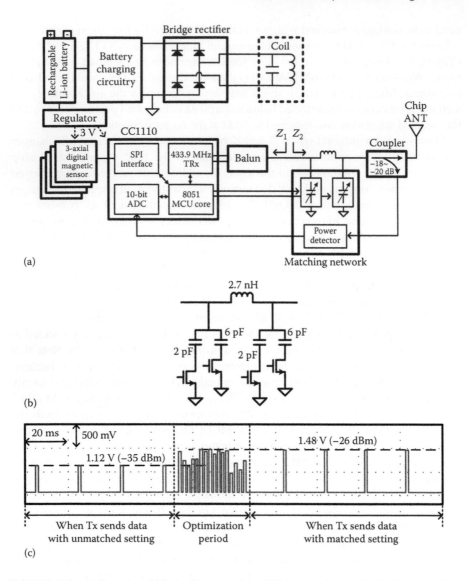

(a)

(b)

(c)

FIGURE 4.5 (a) Functional block diagram of the iTDS operating at 433.9 MHz, (b) CLC pi-network for the adaptive impedance matching inside the mouth, and (c) output voltage of the power detector during the adaptive impedance matching at 433.9 MHz.

from the intraoral environment [68]. In the design of the iTDS-2, we selected a planar inverted-F antenna (PIFA) by considering three important parameters: radiation efficiency, antenna gain (directivity), and polarization [69,70]. Radiation efficiency can be increased by the PIFA because it has a wide radiation surface while it can be folded into a thin structure inside the arch-shaped dental retainer. The PIFA has limited backward radiation, which minimizes the EM power absorption toward the back of the head, while enhancing forward radiation toward the thinnest part of the intraoral surrounding to deliver more power to the TDS-UI receiver. Finally, the PIFA exhibits moderate to high gain in both vertical and horizontal states of polarization, which is also proper for the iTDS where the angle between Tx and Rx antennas changes according to the users' body posture [71–73]. The PIFA was simulated using HFSS software (ANSYS, Canonsburg, Pennsylvania), and an average radiation efficiency of 9.26% was obtained from simulation. The iTDS-2 also incorporates a wounded magnetic coil to employ a near-field communication by inductive coupling via 27 MHz, which can be more efficient than EM wave propagation at short distances. In addition, the human body is almost transparent to the near-field communication at 27 MHz because its wavelength is >10 m and it mainly depends on the magnetic field [74].

4.4 POWER MANAGEMENT

4.4.1 DESIGN CRITERIA OF THE POWER MANAGEMENT BLOCK

To shrink the size of the TDS electronics to comfortably fit inside the mouth, the number of off-chip components and the size of battery should be minimized. An important design objective for the current prototype was to extend the continuous operating time of the iTDS beyond 24 h, such that the users can effectively use it for a full day before a recharge would be necessary. Our solution was to integrate as many blocks as possible on a SoC plus an adjustable power scheduling mechanism to duty cycle the magnetic sensors, readout channels, and Tx to lower the average power consumption of the intraoral appliance [75]. The magnetoresistors on each leg of the HMC1043 sensing bridges change from 0.8 to 1.5 kΩ on the basis of the magnetic field intensity and its orientation, leading to current drains of up to 2.25 mA per bridge at V_{DD} = 1.8 V. Obviously with such a high power consumption, each sensor module should be activated only as long as the AFE is sampling its output. Similarly, the Tx block has a high power consumption and needs to be duty cycled. In our design, the duty cycling ratio of the magnetic sensors and AFE is selectable among 2.3%, 5%, 9%, and 18%, while the Tx is fixed at 25%. According to the TDM control sequences and the duty cycling ratio, the power management integrated circuit (PMIC) turns on/off the supply for each block, with the help of the on-chip digital controller.

The iTDS electronics should be hermetically sealed and carefully packaged in a dental retainer based on every user's unique oral anatomy to be safe and comfortable to wear over extended periods for their daily lives. Similar to implantable microelectronic devices, the hermetic sealing should prevent both leaching of moisture into the iTDS electronics and diffusion of harmful chemicals into the oral space. Proper shaping and fitting of the iTDS dental retainer can be accomplished using well-developed orthodontic methods, starting from a mold that is prepared from the user's dental impression. However, since there is no way for making electric contacts to the hermetically sealed electronics, both recharging of the iTDS Li-ion battery and programming of its configuration register should be accomplished wirelessly via an inductive link that operate at 13.56 MHz. A wire-wound coil, resonating at 13.56 MHz, has been embedded in the iTDS to receive AC power, when it is placed inside the TDS-UI. The same power carrier is amplitude shift keyed (ASK) for data transmission. PMIC rectifies the received AC signal to charge the battery and amplitude demodulates it to recover the configuration data.

4.4.2 WIRELESS CHARGING VIA 13.56 MHz INDUCTIVE LINK

During normal iTDS operation, PMIC is only in charge of power scheduling and bias generation; thus, most of its subblocks are off. However, when the iTDS dental retainer is placed inside the charging cup of the TDS-UI, the 13.56-MHz power carrier couples onto the L_2C_2 tank, generating an AC signal across the full-wave active rectifier inputs, which supplies the rest of PMIC and charges the iTDS embedded 50-mAh Li-ion battery, as shown in Figure 4.6a. The design and operation of the active full-wave rectifier, which offers high AC-to-DC power conversion efficiency

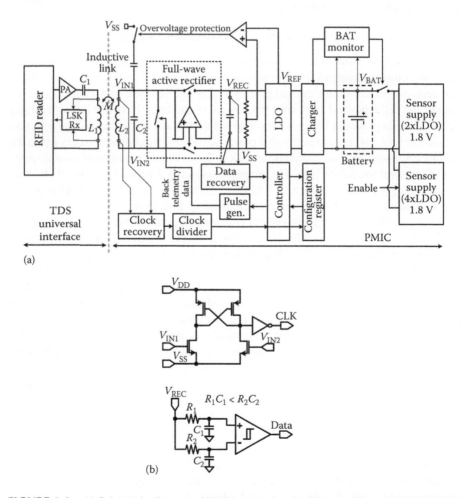

FIGURE 4.6 (a) Schematic diagram of PMIC, including rectifier, regulator, battery charger, and bidirectional data telemetry and (b) clock and data recovery circuits for the forward data telemetry.

in the order of 80% at 13.56 MHz thanks to its offset-controlled high-speed comparators and optimally sized switches, can be found in Ref. [76].

A low dropout regulator follows the rectifier, which provides the battery charger block with a constant 4.4-V supply. The battery charger provides a constant charging current of 6.8 mA to the battery as long as $V_{BAT} < 4.2$ V. When V_{BAT} is charged near 4.2 V, the charger switches from constant current to constant voltage mode and keeps V_{BAT} at 4.2 V to continue charging the battery without damaging it. During constant voltage mode, the charging current gradually decreases, until the charger stops charging the battery when the current goes below 5% of its nominal value.

4.4.3 BIDIRECTIONAL DATA TELEMETRY

The PMIC has bidirectional data telemetry capability with the RFID reader in the TDS-UI that drives the inductive link. Figure 4.6b shows the schematic diagrams of the clock and data recovery circuits for the forward data telemetry. Clock recovery circuit generates the clock signal by comparing the 13.56-MHz sinusoidal signal across the $L_2 C_2$ tank. The recovered clock is then buffered and divided by 256 to provide a 53-kHz master clock signal for the rest of the system. For data recovery, variations on the V_{REC} attributed to ASK of the power carrier by the RFID reader are fed into the data recovery circuit. This simple circuit detects V_{REC} amplitude variations using two paths with different time constants, $R_1 C_1 < R_2 C_2$, which are connected to a hysteresis comparator. The difference between input node voltages after V_{REC} amplitude transitions results in the recovered forward data bit stream at the output of the comparator, which are sampled and delivered to the configuration register by the back telemetry controller, as shown in Figure 4.6a. This circuit also generates a short pulse for every detected bit "1" and applies it to the load shift keying (LSK) mechanism of the active rectifier [76]. Shorting the rectifier input results in a sudden drop in V_{IN} and increased current in L_1. The current and voltage variations in L_1 are detected by the RFID reader and used to recover the LSK back telemetry data.

4.5   SYSTEM IMPLEMENTATION AND EVALUATION

4.5.1 EXTERIOR DESIGN

The first prototype of the iTDS, referred to as the iTDS-1 in this chapter, was implemented as a form of a palatal retainer [12]. Palatal dental retainers locate in the palatal vault space, between the tongue and the palate, a space naturally designated for tongue movements in speech and a temporary reservoir of food during ingestion [77]. Several smart intraoral devices have been implemented as palatal dental retainers, such as TTK, ITCS, and oral electrotactile display [13,17,19]. Figure 4.7a and b show the actual implementation of the TTK and the iTDS-1, occupying the palatal vault space. There are several drawbacks in locating the iTDS-1 in the palatal vault space. First, the iTDS-1 limits the palatal space and impedes proper tongue posture and movement necessary for clear speech [77]. Second, limiting the palatal space potentially degrades the iTDS performance because that space is needed for spacing out the tongue command locations and facilitating their easy selection.

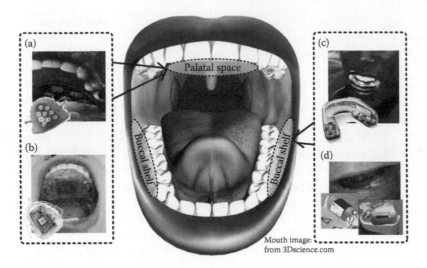

FIGURE 4.7   Intraoral appliances with built-in electronics: (a) TTK, (b) iTDS-1, (c) X2impact mouth guard, and (d) Intellidrug intraoral drug-delivery system.

Intraoral space is also needed for passage of the tongue from one command location to another rapidly and naturally. Therefore, any object on the way might hinder the natural tongue movement. Finally, when the iTDS is worn on the palate, it blocks the palatal tactile feedback that helps users acquire a better sense of the magnetic tracer position and reproduce accurate tongue commands.

To preserve the palatal vault space, a new arch-shaped iTDS, referred to as the iTDS-2 in this chapter, has been designed to be located on the lower jaw, filling the buccal shelf area [61]. The buccal shelf area, a space between molar teeth and inner cheeks, works as a primary stress-bearing area of the mandibular arch and has been used for appliances, such as a mouthpiece or a denture [78]. Figure 4.7c and d show examples of smart intraoral appliances with built-in electronics, designed to be worn on the lower jaw, in the buccal shelf area. We have selected the lower jaw over the upper jaw, to minimize protrusion of the lips and preserve the intraoral space. The intraoral anatomy provides less space between the upper teeth and upper lip than that between the lower teeth and lower lip [77]. Moreover, when the mouth is open, it is easier for the tongue to maintain its relative position with respect to the lower jaw than the upper jaw. This makes it easier to issue the iTDS commands. Nonetheless, a detailed comparative assessment of various intraoral spaces in terms of efficacy and comfort level for a wireless, noncontact tongue-based control surface is necessary.

4.5.2 Size Considerations

The palate is a 3D structure indicating the space at the roof of the mouth surrounded by the concave elliptical bony plate and can be characterized by length, width, and depth, which are defined as end-to-end distances of the palatal space in the anterior–posterior direction, medial–lateral direction, and superior–inferior direction, respectively. From the literature survey, we could find the mean length over the molar teeth as 29 mm, mean width between the first molar teeth as 32 mm, and mean depth as 10 mm. For the palatal retainer production, the boundary of the palate needs to be defined more accurately to not interfere with the original function of the palate [79]. The hamular notch is a key clinical landmark because the vibrating line ("Ah" line), which is used to define the maximum posterior extent of the palatal retainer, runs bilaterally through hamular notches. The vibrating line can also be defined as "the imaginary line across the posterior part of the palate marking the division between the movable and immovable tissues of the soft palate which can be identified when the movable tissues are moving." It is hard to define the maximum thickness of the palatal retainer. However, to minimize the interference to the function of the palatal space as a temporal food reservoir and the support structure for pronunciation, it is important to minimize the thickness of the palatal retainer.

To determine the maximum allowable size of the PCB and electronics for the arch-shaped retainer, we reviewed literature on the buccal shelf area, which is defined as the space bounded on the medial side by the crest of the residual ridge, on the lateral side by the external oblique ridge, in the mesial area by the buccal frenulum, and on the distal side by the masseter muscle [78]. The length of the buccal shelf area is approximately the length of three consecutive molar teeth, reported as ~30 mm, on average [80]. The height of the buccal shelf area can be defined as the total height of molar teeth including both the crown and root lengths. According to Miyabara [81], the heights of the first, second, and third molars were 19.3, 18.4, and 17.2 mm, respectively, in 46 healthy individuals. We considered the data on the thickness of dentures to find information on the depth of the buccal shelf area, which are ~3 mm over the first and second molars and 6.3 mm over the third molar [82]. However, we understand that most of the data are derived from people with dentures and it can differ from the data from people with normal teeth, because the shape of the buccal shelf gradually transforms in the absence of the teeth.

From the above data, each side of the arch-shaped retainer needs to have 28 mm × 14 mm × 2 mm in length, height, and thickness, respectively, for an average adult to wear it with minimum discomfort. Moreover, as shown in Figure 4.8, both control and supply boards on the right and left sides of the retainer, respectively, were tapered on the posterior side, as a trapezoid, considering the shape of the buccal shelf area [78]. The Ortho-Jet crystal layer over the electronics was made thin to minimize the retainer volume while it still provides the mechanical support. The arch-shaped bridge across two boards, which supports two 5.5-mm-wide, 10-wire flat cable interconnects, was designed to be very close to the lower teeth to minimize protrusion toward the lower lips.

Another important consideration, for both the palatal retainer and the arch-shaped retainer, is the difference between the oral anatomy and available oral space in men and women. Moreover, although the above numbers help us in design of the iTDS prototype, we still need to customize the iTDS for each potential user because everyone has a different oral anatomy.

4.5.3 System Implementation and Packaging

Like any other ATs or medical intraoral device designed for extended usage, safety is paramount in both functionality and performance of the iTDS, because it plays

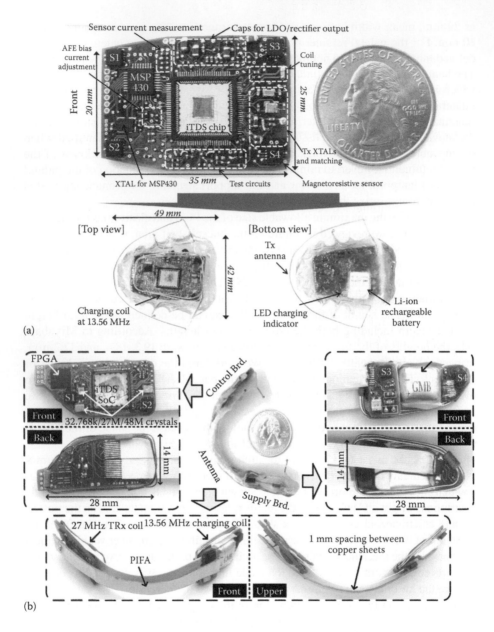

FIGURE 4.8 Actual implementation of the iTDS in the form of (a) a palatal dental retainer and (b) an arch-shaped dental retainer.

such an important role in the users' daily life and contains a rechargeable battery and other electronics. Once the iTDS electronics passes functionality tests, it should be hermetically sealed and carefully packaged in a dental retainer based on every user's unique oral anatomy to be safe and comfortable to wear over extended periods. Similar to implantable microelectronic devices, the hermetic sealing should prevent both leaching of moisture into the iTDS electronics and diffusion of harmful chemicals into the oral space. For example, the iTDS electronics can be coated with biocompatible polymers, such as Parylene C, before being vacuum-molded in acrylic or thermoplastic.

Figure 4.8a and b show the implementation of the iTDS in the form of a palatal dental retainer to be mounted on the upper jaw and an arch-shaped dental retainer to be mounted on the buccal shelf, respectively. Both types of dental retainers are made of orthodontic Ortho-Jet crystal (Lang Dental, Wheeling, Illinois) with two or four stainless still ball clasps (Patterson Dental Supply, St. Paul, Minnesota) according to the oral anatomy and dental composition of users. The Ortho-Jet crystal covers the iTDS electronics and protects them from external forces and the saliva again over the hermetic sealing, while the ball clasps help fix the iTDS onto the teeth. Starting from a design selected by users and doctors and a mold prepared from the user's dental impression, proper shaping and fitting of the iTDS dental retainer can be accomplished using well-developed orthodontic methods. Additional safety and accelerated lifetime measurement tests are also necessary on the strength and the durability of the iTDS package and mechanical structure to avoid the possibility of crack formation or fragmentation under intraoral stress and aging.

4.5.4 INTERFACE WITH OUTSIDE

The custom-designed TDS-UI serves several important purposes in the overall functionality and the usage of the iTDS dental retainer on a regular basis: (1) dual-band receiver (Rx) at 27 and 432 MHz for the raw magnetic sensor data from the iTDS; (2) inductive charger of the iTDS at 13.56 MHz when the dental retainer is placed in the special charging cup (the same inductive link can update the iTDS configuration register via ASK and validate it via LSK); (3) holding the smartphone within the field of view of the iTDS end user either mounted on a PWC or simply on a flat surface; (4) deliver the received raw data from iTDS to the smartphone for applying the SSP algorithm and detecting user commands; (5) establish a reliable hardwired interface between the smartphone and PC/PWC to convert and deliver the TDS navigation commands in the form of USB-compatible digital signals or DC-level adjusted analog signals [36].

Figure 4.9 shows the overall system block diagram of the iTDS-1, to describe how the universal interface interfaces between the iTDS and the outside. The basic operating principle of the iTDS and its SSP algorithm are similar to the eTDS, and the details can be found in Refs. [83,84]. The iTDS dental retainer wirelessly transmits the digitized magnetic sensor raw data packets to a dual-band Rx embedded in the TDS-UI, which in turn delivers them to a smartphone, running the SSP algorithm, to be converted to user-defined tongue commands in real time. Unless the TDS commands are intended for the smartphone, they are sent back to the TDS-UI to be delivered to target devices, which can be either PC or PWC in the current prototype [36].

4.5.5 SYSTEM EVALUATION

After receiving approval from the institutional review board at Georgia Institute of Technology, two healthy subjects (33-year-old male and 30-year-old female)

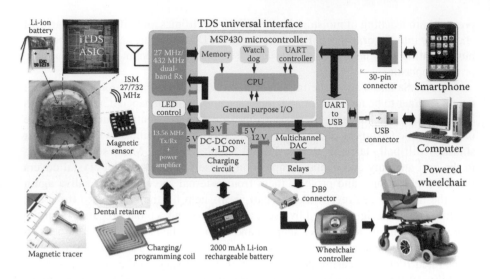

FIGURE 4.9 Overall block diagram of the iTDS-1 including the universal interface to communicate with outside.

participated in the experiment to evaluate the performance of iTDS-1 and iTDS-2. Subjects attached small disk-shaped magnetic tracers (ϕ2.8 mm × 1.0 mm, K&J Magnetics, Jamison, Pennsylvania) on their tongues using dental adhesive and wore their customized palatal and arch-shaped dental retainers [12,61,85]. The iTDS transmits raw magnetic sensor data, changing with the tongue position, to the TDS-UI where the iPod touch identifies one out of seven user-defined commands (left, right, up, down, left-click, right-click, and resting) and sends it to a computer via USB. Subjects trained the system with seven tongue commands and completed three trials for the maze navigation and center-out tapping tasks [86]. Deviation was calculated as the sum of area between the desired path on the maze and the actual traversed path of the mouse cursor on the screen, and completion time is the time taken by the subject to reach the end point. The throughput, measured in bits per second, is an indicator of the rate of information transfer from users to computers via a computer input device. Details of these tasks and their performance measures can be found in Ref. [86].

Table 4.2 shows the summary of the iTDS-1 and iTDS-2 specifications and benchmarking against other intraoral tongue–computer interfaces. The table shows that the iTDS-2 has the longest operating time. The iTDS-2 also has a robust wireless link, thanks to a dual-band Tx that can automatically switch in the presence of interference in each band. Table 4.3 summarizes the performance evaluation results of the iTDS-1 and iTDS-2 in the center-out tapping task and compares them with eTDS, ITCS, Sip 'n' Puff, and computer mouse using results reported in previous publications [12,26,61,87]. Subjects on average demonstrated 22% better performance with the iTDS-2 over the iTDS-1. We speculate that this is because the iTDS-2 does not limit the intraoral space, where the tongue moves, and also preserves the tactile feedback through the palate. The iTDS-2 shows 73% improved performance over the Sip 'n' Puff, which is one of the most popular ATs among those for tetraplegia. The computer mouse still outperforms the iTDS-2 by a factor of 3 to 1. However, this gap

TABLE 4.2
Specifications of iTDS-1, iTDS-2, ITCS, and Optical Tongue Gesture

Specifications		iTDS-1 (12)	iTDS-2 (61)	ITCS (17)	Optical Tongue Gesture (14)
Process		0.5-μm std. CMOS	0.35-μm std. CMOS	Off the shelf	Off the shelf
Die area		3.8×3.7 (mm^2)	2.4×5.0 (mm^2)	–	–
V_{DD}		1.8 (V)	1.8 (V)	–	–
Sensor	Type	Magnetoresistive	Magnetoresistive	Inductive	Infrared proximity
	Channel	12	12	18	4
	Sensitivity	1.8 (mV/gauss)	1.8 (mV/gauss)	–	0.2–0.5 (V/mm)
	Sampling	64 (Hz)	64 (Hz)	30 (Hz)	90 (Hz)
RF	Frequency band	27/432 (MHz)	27/432 (MHz)	2.4 (GHz)	Hardwired
	Switching between bands	Manual	Automatic	Single band	
	Data rate	64 (kbps)	250 (kbps)	–	
Power	P_{avr} (27 MHz)	3.7 (mW)	2.8 (mW)	–	Hardwired
	P_{avr} (432 MHz)	4.7 (mW)	3.3 (mW)		
	Battery type	Li-ion 50 mAh	Li-ion 50 mAh	Li-ion 20 mAh	
	Battery lifetime	17.2 (h), worst case	27.3 (h), worst case	15 (h)	
Prototype	Shape	Palatal shape	Arch shape	Palatal shape	Palatal shape
	Size	$49 \times 42 \times 15$ (mm^3)	$54 \times 28 \times 15$ (mm^3)	$35 \times 25 \times 15$ (mm^3)[a]	$41 \times 38 \times 15$ (mm^3)[a]
	Volume	20 (mL)	12 (mL)	–	–
	Weight	75 (g)	30 (g)	–	–

[a] Estimated from publication.

TABLE 4.3
Performance of the Intraoral Tongue–Computer Interfaces for a Computer Access Task: iTDS-1, iTDS-2, eTDS, ITCS, Sip 'n' Puff, and Computer Mouse

Specifications	iTDS-1 (12)	iTDS-2 (61)	eTDS (26)	ITCS (87)	Sip 'n' Puff (26)	Mouse (26)
Throughput (bits/s)	0.83	1.25	1.48	0.85	0.72	3.66
Number of commands	6	6	6	8	4	–

is expected to narrow as the iTDS users gain more experience and future versions of the iTDS become smaller, while using more power-efficient circuitry. Development of more advanced sensor signal processing algorithms to extract more information from the raw magnetic sensor data, while attenuating external interference and correcting user errors, will also help with this trend.

4.6 CONCLUSION

This chapter presented a design procedure and techniques of a wireless intraoral tongue–computer interface, by following the signal path from the tongue to the external receiver, using the iTDS as a practical example. By employing SoC technology, aggressive duty cycling, inductive wireless charging, and adaptive dual-band wireless transceiver, the whole system of the iTDS has been implemented inside a wearable dental retainer. The iTDS dental retainer works well inside the mouth and transmits sensor data wirelessly to the external receiver for >24 h lifetime without additional recharge.

The iTDS verified its successful operation with two human subjects and showed better performance than the existing ATs, despite several hurdles to work inside the mouth, such as a stringent size constraint and large signal attenuation at the body. Although the size of the iTDS needs to be further reduced for user satisfaction and the wireless connectivity needs to be further tested in harsh environments, prototypes of the iTDS showed a possibility of a wireless intraoral tongue–computer interface to become a strong candidate as an advanced AT. We expect that, in the near future, a wireless intraoral tongue–computer interface as the iTDS will improve the quality of life for severely disabled people.

REFERENCES

1. A.M. Cook and S.M. Hussey. 2007. *Assistive Technologies: Principles and Practice*, 3rd ed. New York: Mosby.
2. T. Hirsch, J. Forlizzi, J. Goetz, J. Stroback, and C. Kurtz. 2000. The ELDer project: Social and emotional factors in the design of eldercare technologies. *Proc. ACM Conf. on Universal Usability*, Arlington, VA, USA, pp. 72–79, Nov.
3. Adaptive Switch Labs, Inc. Available at http://www.asl-inc.com/catalog.
4. Origin Instrument: Sip/puff switch. Available at http://www.orin.com/access/sip_puff /index.htm.
5. M.R. Williams and R.F. Kirsch. 2008. Evaluation of head orientation and neck muscle EMG signals as command inputs to a human–computer interface for individuals with high tetraplegia. *IEEE Trans. Neural Syst. Rehabil. Eng.*, vol. 16, no. 5, pp. 485–496, Oct.
6. NaturalPoint, Inc. Available at http://www.naturalpoint.com/smartnav/.
7. Nuance Communications, Inc. Available at http://www.nuance.com/dragon/index.htm.
8. J.R. Wolpaw, N. Birbaumer, D.J. McFarland, G. Pfurtscheller, and T.M. Vaughan. 2002. Brain–computer interfaces for communication and control. *Clin. Neurophysiol.*, vol. 113, pp. 767–791, June.
9. R. Barea, L. Boquete, M. Mazo, and E. Lopez. 2002. System for assisted mobility using eye movements based on electrooculography. *IEEE Trans. Rehabil. Eng.*, vol. 10, no. 4, pp. 209–218, Dec.
10. M. Ghovanloo. 2007. Tongue operated assistive technologies. *Proc. International Conf. IEEE Engineering in Medicine and Biology Society*, Lyon, France, pp. 4376–4379, Aug.
11. E.R. Kandel, J.H. Schwartz, and T.M. Jessell. 1991. *Principles of Neural Science*, 4th ed.: New York: McGraw-Hill, pp. 337–348.
12. H. Park, M. Kiani, H. Lee, J. Kim, J. Block, B. Gosselin, and M. Ghovanloo. 2012. A wireless magnetoresistive sensing system for an intraoral tongue–computer interface. *IEEE Trans. Biomed. Circuits Syst.*, vol. 6, no. 6, pp. 571–585, Dec.

13. M. Giraldi. 1997. Independence day: Tongue-touch controls give Ben a more satisfying self-sufficient lifestyle. *Teamrehab Rep. Mag.*, pp. 14–17.
14. T.S. Saponas, D. Kelly, B.A. Parviz, and D.S. Tan. 2009. Optically sensing tongue gestures for computer input. *Proc. ACM Symposium on User Interface Software and Technology*, Victoria, BC, Canada, pp. 177–180, Oct.
15. C. Salem, and S. Zhai. 1997. An isometric tongue pointing device. *Proc. CHI 97: Human Factors in Computing Systems*, San Jose, CA, USA, pp. 22–27, May.
16. W. Nutt, C. Arlanch, S. Nigg, and G. Staufert, 1998. Tongue-mouse for quadriplegics. *J. Micromech. Microeng.*, vol. 8, no. 2, pp. 155–157, June.
17. L.N.S. Andreasen Struijk, E.R. Lontis, B. Bentsen, H.V. Christensen, H.A. Caltenco, and M.E. Lund. 2009. Fully integrated wireless inductive tongue computer interface for disabled people. *Proc. International Conf. IEEE Engineering in Medicine and Biology Society*, Minneapolis, MN, USA, pp. 547–550, Sept.
18. N.P. Solomon. 2004. Assessment of tongue weakness and fatigue. *Int. J. Orofac. Mycol.*, vol. 30, pp. 8–19, Nov.
19. H. Tang and D.J. Beebe. 2006. An oral tactile interface for blind navigation. *IEEE Trans. Neural Syst. Rehab. Eng.*, vol. 14, no. 1, pp. 116–123, Mar.
20. Y. Danilov, M. Tyler, K.A. Kaczmarek, and C. Luzzio. 2009. Rehabilitation of multiple sclerosis symptoms using cranial nerve non-invasive neuromodulation (CN-NINM). *Poster Presentation in Conf. Society for Neuroscience*, Oct. Available at https://tcnl .bme.wisc.edu/sites/default/files/2009-Danilov-SfN-MS-Poster.pdf.
21. I. Goker and M. Ozkan. 2004. Design of intraoral artificial larynx. *J. Electrical and Electronics Engineering*, vol. 4, no. 2, pp. 1171–1176, June.
22. T. Gottsche and A. Woff. 2006. IntelliDrug—An integrated intelligent oral drug delivery system. *Integrated Microsystems for Biomedicine*, pp. 9–10.
23. X. Huo, J. Wang, and M. Ghovanloo. 2008. A magneto-inductive sensor-based wireless tongue–computer interface. *IEEE Trans. Neural Syst. Rehabil Eng.*, vol. 16, pp. 497–504, Oct.
24. H. Park, J. Kim, X. Huo, I. Hwang, and M. Ghovanloo. 2011. New ergonomic headset for tongue-drive system with wireless smartphone interface. *Proc. Int. Conf. IEEE Engineering in Medicine and Biology Society*, Boston, MA, USA, pp. 7344–7347, Aug.
25. J. Kim, X. Huo, J. Minocha, J. Holbrook, A. Laumann, and M. Ghovanloo. 2012. Evaluation of a smartphone platform as a wireless interface between tongue drive system and electric-powered wheelchairs. *IEEE Trans. Biomed. Eng.*, vol. 59, no. 6, pp. 1787–1796, June.
26. J. Kim, H. Park, J. Bruce, E. Sutton, D. Rowles, D. Pucci, J. Holbrook, J. Minocha, B. Nardone, D. West, A. Laumann, E. Roth, M. Jones, E. Veledar, and M. Ghovanloo. 2013. The tongue enables computer and wheelchair control for people with spinal cord injury. *Sci. Transl. Med.*, vol. 5, no. 213, p. 213ra166, Nov.
27. J. Kim, H. Park, J. Bruce, E. Sutton, D. Rowles, D. Pucci, J. Holbrook, J. Minocha, B. Nardone, D. West, A. Laumann, E. Roth, M. Jones, E. Veledar, and M. Ghovanloo. 2014. The qualitative assessment of tongue drive system by people with high-level spinal cord injuries. *J. Rehabil. Res.*, vol. 51, no. 3, pp. 451–466, July.
28. F. Hannon. 2007. *Literature Review on Attitudes towards Disability*. Disability Research Series 9, Dublin, Ireland: National Disability Authority.
29. H. Park, J. Kim, and M. Ghovanloo. 2012. Development and preliminary evaluation of an intraoral tongue drive system. *Proc. Int. Conf. IEEE Engineering in Medicine and Biology Society*, San Diego, CA, USA, pp. 1157–1160, Sept.
30. H. Istance, R. Bates, A. Hyrskykari, and S. Vickers. 2008. Snap clutch, a moded approach to solving the Midas touch problem. *Proc. Symposium on Eye Tracking Research & Applications*, pp. 221–228, Mar.

31. M.J. Caruso and C.H. Smith. A new perspective on magnetic field sensing, Honeywell Applications Note, AN-201. Available at http://www51.honeywell.com/.
32. I. Vieira, M. Martins, and J. Parracho. Magnetoresistive sensors for a magnetometer. Available at https://escies.org/.
33. D.R. Haynor, C.P. Somogyi, and R.N. Golden. 2000. System and method to determine the location and orientation of an indwelling medical device. U.S. Patent 6 129 668, Oct. 10.
34. A.N. Johnson, X. Huo, M. Ghovanloo, and M. Shinohara. 2012. Dual-task motor performance with a tongue-operated assistive technology compared with hand operations. *J. Neuroeng. Rehab.*, vol. 9, pp. 1–16, Jan.
35. B. Yousefi, X. Huo, J. Kim, E. Veledar, and M. Ghovanloo. 2012. Quantitative and comparative assessment of learning in a tongue-operated computer input device—Part II: Navigation tasks. *IEEE Trans. Info. Tech. in Biomedicine*, vol. 16, no. 4, pp. 633–643, July.
36. J. Kim, H. Park, J. Bruce, D. Rowles, J. Holbrook, B. Nardone, D. West, A. Laumann, E. Roth, and M. Ghovanloo. 2015. Assessment of the tongue-drive system using a computer, a smartphone, and a powered-wheelchair by people with tetraplegia. *IEEE Trans. Neural Sys. and Rehab. Eng.*, Feb.
37. B. Gosselin and M. Ghovanloo. 2011. A high-performance analog front-end for an intraoral tongue-operated assistive technology. *Proc. IEEE Int. Symp. on Circuits and Systems*, pp. 2613–2616, May.
38. J.F. Witte, J.H. Huijsing, and K. Makinwa. 2008. A current-feedback instrumentation amplifier with 5 µV offset for bidirectional high-side current-sensing. *IEEE J. Solid-State Circuits*, vol. 43, pp. 2769–2775, Dec.
39. H. Park and M. Ghovanloo. 2013. A 13-bit noise shaping SAR-ADC with dual-polarity digital calibration. *Analog Integr. Circuits Signal Process.*, vol. 75, no. 3, pp. 459–465, June.
40. H. Lee, D.A. Hodges, and P.R. Grey. 1984. A self-calibrating 15 bit CMOS A/D converter. *J. Solid State Circuits*, vol. SC-19, no. 6, pp. 813–819, Dec.
41. J. Li and U. Moon. 2003. Background calibration techniques for multistage pipelined ADCs with digital redundancy. *IEEE Trans. Circuits Syst. II*, vol. 50, no. 9, pp. 531–538, Sept.
42. K. Kim, S. Yun, S. Lee, and S. Nam. 2012. Low power CMOS super-regenerative receiver with a digitally self-quenching loop. *IEEE Microw. Wirel. Compon. Lett.*, vol. 22, no. 9, pp. 486–488, Sept.
43. N. Cho, J. Yoo, S, Song, J. Lee, S. Jeon, and H. Yoo. 2007. The human body characteristics as a signal transmission medium for intrabody communication. *IEEE Trans. Microwave Theory and Techn.*, vol. 55, no. 5, pp. 1080–1086, May.
44. J. Bae, H. Cho, K. Song, H. Lee, and H. Yoo. 2012. The signal transmission mechanism on the surface of human body for body channel communication. *IEEE Trans. Microwave Theory and Techn.*, vol. 60, no. 3, pp. 582–593, Mar.
45. M.S. Wegmueller, S. Huclova, J. Froehlich, M. Oberle, N. Felber, N. Kuster, and W. Fichtner. 2009. Galvanic coupling enabling wireless implantable communications. *IEEE Trans. Instrumentation and Measurement*, vol. 58, no. 8, pp. 2618–2625, Aug.
46. J.E. Ferguson and D. Redish. 2011. Wireless communication with implantable medical devices using the conductive properties of the body. *Special Report on Expert-Reviews*, vol. 8, no. 4, pp. 427–433, July.
47. K. Gosalia, G. Lazzi, and M. Humayun. 2004. Investigation of a microwave data telemetry link for a retinal prosthesis. *IEEE Trans. Microwave Theory and Techn.*, vol. 52, no. 8, pp. 1925–1933, Aug.
48. J. Kim and Y. Rahmat-Samii. 2004. Implanted antennas inside a human body: Simulations, designs, and characterizations. *IEEE Trans. Microwave Theory and Techn.*, vol. 52, no. 8, pp. 1934–1943, Aug.

49. R. Warty, M.R. Tofighi, U. Kawoos, and A. Rosen. 2008. Characterization of implantable antennas for intracranial pressure monitoring: Reflection by and transmission through a scalp phantom. *IEEE Trans. Microwave Theory and Techn.*, vol. 56, no. 10, pp. 2366–2376, Oct.

50. T. Karacolak, A.Z. Hood, and E. Topsakal. 2008. Design of a dual-band implantable antenna and development of skin mimicking gels for continuous glucose monitoring. *IEEE Trans. Microwave Theory and Techn.*, vol. 56, no. 4, pp. 1001–1008, Apr.

51. Z.N. Chen, G.C. Liu, and T. See. 2009. Transmission of RF signals between MICS loop antennas in free space and implanted in the human head. *IEEE Trans. Antennas and Propagation*, vol. 57, no. 6, pp. 1850–1854, June.

52. L. Xu, M.Q.-H. Meng, H. Ren, and Y. Chan. 2009. Radiation characteristics of ingestible wireless devices in human intestine following radio frequency exposure at 430, 800, 1200, and 2400 MHz. *IEEE Trans. Antennas and Propagation*, vol. 57, no. 8, pp. 2418–2428, Aug.

53. C.A. Roopnariane, M. Tofighi, and C.M. Collins. 2010. Radiation performance of small ingestible antennas in head at MICS, ISM, and GPS bands. *Proc. IEEE Conf. Northeast Bioengineering*, pp. 1–2, Mar.

54. A. Kiourti and K.S. Nikita. 2012. Miniature scalp-implantable antennas for telemetry in the MICS and ISM bands-design, safety considerations and link budget analysis. *IEEE Trans. Antennas and Propagation*, vol. 60, no. 8, pp. 3568–3575, Aug.

55. M. Ghovanloo and K. Najafi. 2007. A wireless implantable multichannel microstimulating system-on-a-chip with modular architecture. *IEEE Trans. Neural Systems and Rehab. Eng.*, vol. 15, no. 3, pp. 449–457, Sept.

56. A.J. Johansson. 2004. *Wireless Communication with Medical Implants: Antennas and Propagation.* Ph.D. Thesis, Lund University, Switzerland.

57. D. Pozar. 2011. *Microwave Engineering*, 4th ed., New York: John Wiley and Sons.

58. S. Gabriel, R.W. Lau, and C. Gabriel. 1996. The dielectric properties of biological tissues: II. Measurements in the frequency range 10 Hz to 20 GHz. *Phys. Med. Biol.*, vol. 41, no. 11, pp. 2251–2269, Nov.

59. F. Hatmi, M. Grzeskowiak, T. Alves, S. Protat, and O. Picon. 2011. Magnetic loop antenna for wireless capsule endoscopy inside the human body operating at 315 MHz: Near field behavior. *Proc. IEEE Mediterranean Microwave Symposium*, pp. 81–87, Nov.

60. H. Park and M. Ghovanloo. 2014. Wireless communication of intraoral devices and its optimal frequency selection. *IEEE Trans. Microwave Theory and Techn.*, vol. 62, no. 12, pp. 3205–3215, Dec.

61. H. Park and M. Ghovanloo. 2014. An arch-shaped intraoral tongue drive system with built-in tongue–computer interfacing SoC. *Sensors*, vol. 14, no. 11, pp. 21565–21587, Nov.

62. S. Rai, J. Holleman, J. Pandey, F. Zhang, and B. Otis. 2009. A 500 µW neural tag with 2 µVrms AFE and frequency-multiplying MICS/ISM FSK transmitter. *IEEE ISSCC Dig. Tech. Papers*, Feb.

63. H. Schantz and J. Fluhler. 2006. Near-field technology—An emerging RF discipline. *Proc. European Conf. Antennas and Propagation*, keynote address, Nov.

64. P. Sjöblom and H. Sjöland. 2005. An adaptive impedance tuning CMOS circuit for ISM 2.4-GHz band. *IEEE Trans. Circuits and Systems-I*, vol. 52, no. 6, pp. 1115–1124, June.

65. J. Mingo, A. Valdovinos, A. Crespo, D. Navarro, and P. García. 2004. An RF electronically controlled impedance tuning network design and its application to an antenna input impedance automatic matching system. *IEEE Trans. Microwave Theory and Techn.*, vol. 52, no. 2, pp. 489–497, Feb.

66. J. Kim, X. Huo, J. Minocha, J. Holbrook, A. Laumann, and M. Ghovanloo. 2012. Evaluation of a smartphone platform as a wireless interface between tongue drive system and electric-powered wheelchairs. *IEEE Trans. Biomed. Eng.*, vol. 59, no. 6, pp. 1787–1796, June.

67. C. Capps. 2001. Near field or far field? Santa Monica, CA, USA: *EDN Mag.*, pp. 95–102.
68. C.L. Yang, C.L. Tsai, and S.H. Chen. 2013. Implantable high-gain dental antennas for minimally invasive biomedical devices. *IEEE Trans. Antennas Propag.*, vol. 61, no. 5, pp. 2380–2387, May.
69. J. McLean, R. Sutton, and R. Hoffman. *Interpreting antenna performance parameters for EMC applications—Part I: Radiation efficiency and input impedance match.* Available at http://tdkrfsolutions.com/images/uploads/brochures/antenna_paper_part1 .pdf.
70. J. McLean, R. Sutton, and R. Hoffman. *Interpreting antenna performance parameters for EMC applications—Part II: Radiation pattern, gain, and directivity.* Available at http://tdkrfsolutions.com/images/uploads/brochures/antenna_paper_part2.pdf.
71. P.M. Evjen and G.E. Jonsrud. *AN003: SRD Antennas. Texas Instrument Applications Note.* Available at http://www.ti.com/lit/an/swra088/swra088.pdf.
72. Y. Rahmat-Samii, and J. Kim. 2006. *Implanted Antennas in Medical Wireless Communications*, 1st ed. San Rafael, CA, USA: Morgan & Claypool.
73. Z. Zhang. 2011. *Antenna Design for Mobile Devices*, 1st ed. New York: John Wiley & Sons.
74. F. Hatmi, M. Grzeskowiak, T. Alves, S. Protat, and O. Picon. 2011. Magnetic loop antenna for wireless capsule endoscopy inside the human body operating at 315 MHz: Near field behavior. *Proc. IEEE Mediterranean Microwave Symposium*, pp. 81–87, Nov.
75. S. Cho and A.P. Chandrakasan. 2001. Energy efficient protocols for low duty cycle wireless microsensor networks. *Proc. IEEE Int. Conf. Acoustics, Speech, and Signal Processing*, pp. 2041–2044, May.
76. H. Lee and M. Ghovanloo. 2011. An integrated power-efficient active rectifier with offset-controlled high speed comparators for inductively-powered applications. *IEEE Trans. Circuits Syst. I*, vol. 58, no. 8, pp. 1749–1760, Aug.
77. J.C. Posnick. 2013. *Principles and Practice of Orthognathic Surgery*, 1st ed. New York: Elsevier, pp. 227–263.
78. R. Arthur, J.R. Ivanhoe, and K.D. Plummer. 2009. *Textbook of Complete Dentures*, 6th ed. Shelton, CT, USA: PMPH-USA, pp. 34–43.
79. W.B. Mushtaha. Introduction in prosthodontics (dental prosthetics). Available at http://www.up.edu.ps/ocw/repositories/pdf-archive/MGDS2105.101_24022009.pdf, Accessed October 13, 2015.
80. S.A. Fernandes, F. Vellini-Ferreira, H. Scavone-Junior, and R.I. Ferreira. 2011. Crown dimensions and proximal enamel thickness of mandibular second bicuspids. *Braz. Oral Res.*, vol. 25, no. 4, pp. 324–330, July.
81. T. Miyabara. 1916. An anthropological study of the masticatory system in the Japanese: The teeth. *Dent. Cosmo*, vol. 58, no. 7, pp. 739–749, July.
82. J. He, T. Chou, H. Chang, J. Chen, Y. Yang, and J. Moore. 2007. Predictable reproduction of the Buccal shelf area in mandibular dentures. *Int. J. Prosthodont.*, vol. 20, no. 5, pp. 535–537, Sept.
83. X. Huo and M. Ghovanloo. 2009. Using unconstrained tongue motion as an alternative control surface for wheeled mobility. *IEEE Trans. on Biomed. Eng.*, vol. 56, no. 6, pp. 1719–1726, June.
84. E.B. Sadeghian, X. Huo, and M. Ghovanloo. 2011. Command detection and classification in tongue drive assistive technology. *Proc. Int. Conf. IEEE Engineering in Medicine and Biology Society*, pp. 5465–5468, Sept.
85. H. Park, J. Kim, and M. Ghovanloo. 2012. Intraoral tongue drive system demonstration. *Proc. International Conf. IEEE Biomedical Circuits and Systems*, Hsinchu, Taiwan, p. 81, Nov.

86. B. Yousefi, X. Huo, E. Veledar, and M. Ghovanloo. 2011. Quantitative and comparative assessment of learning in a tongue-operated computer input device. *IEEE Trans. Info. Tech. Biomed.*, vol. 15, no. 5, pp. 747–757, June.
87. H.A. Caltenco, B. Björn, and L.N. Struijk. 2014. On the tip of the tongue: Learning typing and pointing with an intra-oral computer interface. *Disability and Rehabilitation. Assist. Technol.*, vol. 9, no. 4, pp. 307–317, July.

5 Energy-Efficient Hierarchical Wireless Sensor Networks Based on Wake-Up Receiver Usage

Heikki Karvonen and Juha Petäjäjärvi

CONTENTS

5.1 INTRODUCTION

Energy-efficient communication is crucial in wireless sensor network (WSN), which can have many effective applications in the medical field, in order to enable that the WSNs can be deployed for long periods without battery replacement or charging. Typically, wireless transceiver consumes most of the sensor nodes' energy resources [1–7]; therefore, a careful design must be performed for communication techniques and protocols. Indeed, a huge amount of research work has been carried out during the past decade to improve the energy efficiency by optimizing protocols using a layered and cross-layer approach.

This chapter will be focused on intelligent hierarchical WSN architecture, which can be used to effectively utilize heterogeneous devices collecting different types of sensor data from the patient's body or environment, performing autonomous networking, and providing data for the databases of the Internet of Things (IoT). Hierarchical architecture will be introduced in Section 5.2.

In the hierarchical network case, the energy consumption can be decreased by utilizing a wake-up concept that enables to keep the most power-consuming devices at a sleep mode as long as possible. There are two different types of concepts that can be used to enable the wake-up between different hierarchical layers of the architecture: duty cycling–based radios and wake-up receiver (WUR) usage. The wake-up concept needs a joint design of physical and medium access control (MAC) layers. The wake-up concept design issues will be discussed in Section 5.2.2, and a generic wake-up radio–based MAC (GWR-MAC) protocol will be introduced in Section 5.2.3.

Different types of state-of-the-art WUR solutions that can be used to enable the wake-up concept will be introduced, and their future research challenges are outlined in Section 5.3. Energy efficiency comparison results for a GWR-MAC–based hierarchical WSN architecture and conventional duty cycle MAC–based WSN are introduced in Section 5.4 to show that the WUR-based networking has remarkable potential to improve energy efficiency particularly in the target scenarios where events occur rarely. WSN devices' long lifetime will make the applications more user-friendly and therefore it will foster widespread deployments. When the number of WSNs providing sensor data for the databases of the IoT can be increased, the possibilities for horizontal deployment of different types of applications will be enormous. It has been estimated that 50 billion devices and objects will be connected to the IoT by 2020 [8]. As introduced in this book, wireless medical networks have enormous possibilities to improve quality and effectiveness of healthcare. Examples of envisaged applications that can be built by utilizing the described intelligent hierarchical architecture will also be outlined in Section 5.4.

5.2   HIERARCHICAL ARCHITECTURE

The WSN architectures can be divided roughly into two categories: flat and hierarchical. In the flat architecture case, all the network nodes are at the same level and they have similar roles from the communication point of view and typically also similar characteristics. In the hierarchical network case, the nodes have different characteristics and roles at different hierarchical layers. The flat network structure is simpler but it cannot provide efficient communication, especially when the network is composed of a large amount of nodes. The hierarchical network structure has been found to provide more efficient communication in the case of heterogeneous networks since the nodes' operations can be designed so that the overall performance will be improved in comparison with a flat architecture.

At the early stage of the WSN research, the target applications included a homogeneous set of sensor devices, which were performing a simple sensing task and reporting sensor observations to a central (sink) node. That star topology is still valid for many WSN applications. However, the development has led to an emergence of heterogeneous networks, which include different types of devices with varying capabilities, enabling the implementation of more versatile application scenarios. Support for heterogeneous devices is needed in WSNs for energy efficiency, scalability, and quality of service purposes. The network of a heterogeneous set of devices must be designed carefully to enable efficient and reliable operation. The previously proposed architectures can be divided into intranetwork and in-network approaches. In the intranetwork approach, the main design goal is efficient communication between the WSN and backbone (see, e.g., Refs. [9] and [10]). The in-network approaches of WSNs typically limit to two-tier topologies, where the higher tier (gateway) collects data and forms connection to a backbone network. The lower tier nodes can be simpler and they save energy by communicating directly, in a star-topology fashion, only with the gateway [11]. Several WSN protocols, for example, ZigBee [12] and Z-Wave [13], make a distinction between a routing (full-function device) and a nonrouting (reduced-function device) device. However, that approach also allows clustering of nodes and designating only one node at a time as an energy-consuming higher tier node (cluster head) [14,15]. The multitier architectures are typically designed for a specific application, for example, hospital environment [16], traffic monitoring system [17], surveillance [18], environmental monitoring [19], smart home [20], or underwater acoustic sensor networks [21].

The hierarchical network's total energy consumption can be decreased by taking carefully into account the characteristics of the heterogeneous devices at different layers of the architecture. Different types of devices' functionalities must be designed so that communication and sensing requirements can be met while maintaining the low energy consumption. From a communication's energy consumption point of view, it is important to maximize the length of the nodes' sleep mode and minimize the number of retransmissions. That is particularly important in a heterogeneous network for the higher tier nodes, which are the most power consuming.

Intelligent hierarchical architecture for heterogeneous WSNs will be introduced in Section 5.2.1. To enable energy efficiency by putting the nodes into the sleep mode, the heterogeneous hierarchical network requires a method for awakening the nodes when required from the application point of view. For that purpose, duty cycle–based radio is the traditional approach while WURs have started to gain more and more research attention in recent years. Both approaches will be discussed in Section 5.2.2. A GWR-MAC protocol will be introduced in Section 5.2.3.

5.2.1 Architecture for a Heterogeneous Network

Network architecture based on hierarchical levels of intelligence and usage of wake-up concept has been introduced [22]. It can be used in various WSN and wireless body area network (WBAN) application scenarios, which include different types of devices. The wake-up functionality is seen as a long lifetime enabler, particularly for the wireless long-lasting (e.g., medical, surveillance, structural, environmental, and industrial) monitoring systems, which have many possible application scenarios in both private and public sectors. Because of a very wide application space, the hierarchical architecture is designed to be flexible and scalable to varying configurations [22]. Therefore, a high-level architecture that is independent of the specific implementation techniques (e.g., radio interfaces) is defined in order to enable that different application developers have flexibility to choose the most suitable implementation technique. The proposed high-level architecture can be used as a starting point for the network design. In the discussed hierarchical network case, the nodes are categorized into different architectural layers on the basis of their capabilities and functionalities. The high-level architecture and the functionalities offered at the different layers are illustrated in Figure 5.1.

In this design, the hierarchical layers of the architecture are named to be meaningful as follows: elementary layer (EL), intermediate layer (IL), advanced layer (AL), and outer layer (OL). The complexity, cost, capabilities, and performance of the devices at different layers increase from the bottom to the top. Consequently, the devices at the lowest layer consume less energy than the devices at the higher layers. The objective of the intelligent hierarchical design is to decrease the overall energy consumption by using the low-complexity devices for continuous event monitoring and data collection while keeping the more power-consuming higher-layer devices in the sleep mode as long as possible. The EL nodes will wake up the higher-complexity devices only when required. The EL devices are usually simple sensor nodes providing basic sensing and possibly also actuation services. However, the devices at the lowest layer are elementary from the application service point of view since they provide the essential data about the monitored event or object. The networking services offered by the EL nodes are communication with the devices at the IL, and in the mesh network case, the EL sensor nodes can communicate also with each other. The simplest service that the EL node can offer is just to send a simple message containing, for example, a body temperature or environment humidity value sensed by the node. The IL nodes have more capabilities, and they can offer higher-performance functionalities and sensing, such as performing electrocardiography or recording a video. The IL nodes may also perform data aggregation for information

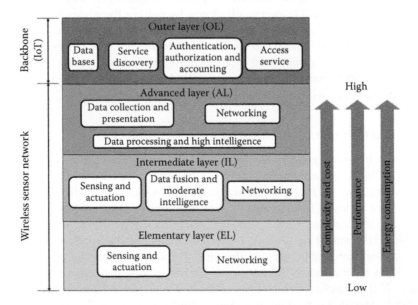

FIGURE 5.1 High-level architecture for a hierarchical network with heterogeneous devices.

collected from EL nodes and make decisions on the basis of the data by using, for example, pattern recognition algorithms. Therefore, the IL nodes can decide whether a sensed event is so critical that also the AL needs to be awakened. The IL nodes offer important networking services by communicating with both upper- and lower-layer devices. The AL nodes are the most intelligent devices in the architecture, and they will eventually collect all the relevant data from the lower layers and make intelligent decisions, process data, and act as a gateway between the WSN and the backbone network. Therefore, AL devices must provide adequate service interfaces so that the application data can be offered to the end user through backbone. The OL is the backbone (public or private) infrastructure providing wireless or wired communication (e.g., broadband or cellular network) back-end systems and application servers. Since this layer is not part of the actual WSN architecture, it is called OL in this design.

Figure 5.2 illustrates an example of application scenarios that have been implemented (in the WAS-project funded by the Finnish Funding Agency for Innovation [Tekes]), using the proposed hierarchical architecture principle. Subnetwork 1 illustrates a WSN where the sensor nodes (TelosB [23]) collect data from the office environment and send it to the embedded computer (FriendlyARM [24]), which forwards the data to the desktop computer with an Ethernet connection to the backbone network (Internet). Subnetwork 2 illustrates a surveillance WSN where the sensor nodes (TUTWSN [25]) detect movements in the monitored area and wake up a wireless local area network (WLAN) camera node to record a video once the moving object has been sensed. The WLAN camera node sends the video to an embedded computer, which forwards the data to the backbone network (Internet). An authenticated end user can access the data from different subnetworks through the Internet and

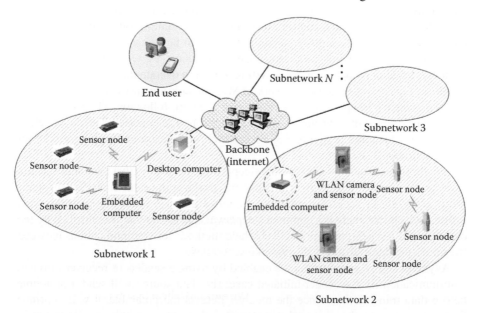

FIGURE 5.2 Distributed heterogeneous network example.

receive, for example, automatic alarms about intruders in the monitored area. In medical applications, this type of surveillance network with tailored sensors could, for example, monitor elderly or Parkinson disease patients at a home environment and provide alarms for the patient itself and for the nursing staff at the hospital when movements are detected in unusual or dangerous areas. The patient's WBAN can also be implemented using the hierarchical architecture principle. The information collected by WBAN would provide useful information about the patient state, for example, when the surveillance network has triggered the alarm and caught medical persons to pay attention to the patient because of the detected event (DE).

5.2.2   WAKE-UP CONCEPT

As was described, the idea of the energy-efficient hierarchical architecture is to keep the most power-consuming nodes in a sleep mode as long as possible. For that purpose, there is a need for a low-power wake-up concept, which can be implemented by using a specific WUR or by using a duty-cycled MAC protocol.

The design of the wake-up concept must take into account physical and MAC layer characteristics. There are different types of MAC protocols that try to improve energy efficiency. Their efficiency depends on the application characteristics since different types of networks require different types of solutions. Indeed, the scalability and adaptability to network changes are important MAC protocol design objectives because in that way the protocol performance can be ensured in many types of scenarios. MAC protocols must be designed to take care of the packet collision avoidance, idle listening, and overhearing with minimum control overhead. Packet collisions must be avoided to keep the number of retransmissions low. Idle listening occurs when radios listen to the channel redundantly, when there are no incoming transmissions. Retransmissions and control overhead decrease data throughput and increase energy consumption because redundant bits need to be transmitted. Consequently, the available sleep time of the sensor nodes also decreases. Overhearing should be avoided so that only the target nodes will receive and decode the packets.

Most of the proposed sensor network MAC protocols are duty cycle based; that is, the radios have a sleep/awake scheduling that they are following. The MAC protocols can be divided into synchronous and asynchronous categories. The duty cycling principle is illustrated in Figure 5.3a for a synchronous case and in Figure 5.3b for an asynchronous case.

Synchronous protocols schedule the sleep/awake periods so that the nodes that are expected to communicate with each other are awake at the same time. Synchronous protocols typically require a centralized control and use clustering of nodes, and inside each cluster, there is a common sleep/awake schedule that is controlled by the cluster head. Asynchronous protocols include methods for communications between the nodes that have different sleep/awake schedules.

Asynchronous communication is enabled by using a sender- or receiver-initiated communication. In the sender-initiated case, the data source will send a preamble before data transmission. Once the receiver detects the preamble, it will continue listening to receive the data packet that will follow the preamble, as illustrated in Figure 5.3b. If the receiver does not detect a preamble, it will go back to sleep mode. In

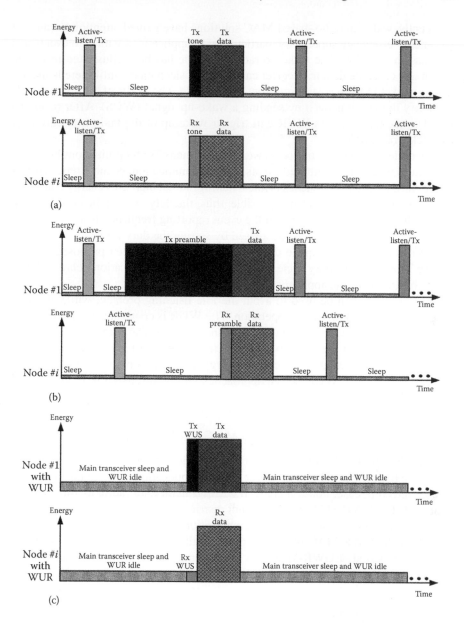

FIGURE 5.3   Principle for (a) synchronous duty cycling, (b) asynchronous duty cycling, and (c) wake-up radio–based MAC.

the receiver-initiated case, the receiver will use probing to query for potential transmissions. In each case, the idle listening will occur if there is no incoming transmission when the nodes wake up to listen to the channel according their schedule. Idle listening is expensive from the power consumption point of view because the transceivers should be in the sleep mode as much as possible in order to save power.

Recently, wake-up radio–based MAC solutions have gained attention because of their energy efficiency superiority, particularly in applications with rare events and transmissions [26–32]. The wake-up radio principle has been illustrated in Figure 5.3c. In this case, the data transceiver can be in the sleep mode until there is incoming data packet to be received from some other node. The WUR of the source mode will notify the target node(s) by sending a wake-up signal (WUS). After receiving the WUS, the target node's WUR will trigger wake-up of the main transceiver for data packet reception.

Duty cycle–based MAC protocols work well in many WSN applications but not in applications where the monitored events, and communication, occur rarely. A drawback of the duty cycle approaches is idle listening, which increases energy consumption and should thus be avoided if possible; thus, the duty cycle radios (DCRs) will listen to the channel unnecessarily if the event reporting frequency is lower than the duty cycle. Idle listening can be decreased by setting the duty cycle very low when the traffic load is low. Adaptive duty cycle protocols have been proposed for that purpose. However, low duty cycle will increase the communication delay and may not be able to satisfy the application requirements.

WURs [28–36] can be used to avoid the idle listening problem while allowing energy-efficient and low-latency operation. The WUR is continuously able to detect the WUS when it is in the ultralow-power standby mode. The communication delay can be avoided when the event occurs and there are data to transmit. Therefore, WURs have the potential to decrease energy consumption in comparison with DCR-based networks. WUR design approaches will be discussed in more detail in Section 5.3.

5.2.3 A Generic-Level WUR-Based MAC for Hierarchical Architecture

A GWR-MAC protocol, which is based on dual-radio approach, is introduced here to enable idle listening avoidance in sensor network applications [30]. In the dual-radio node architecture case, nodes include WUR and main data radio, as illustrated in Figure 5.4. The GWR-MAC protocol is not restricted to any specific WUR technology or data radio technology. Two different options for the wake-up procedure are defined for the GWR-MAC protocol: source initiated and sink initiated. The data transmission period of GWR-MAC can be implemented by using different types of channel access methods, as will be explained below.

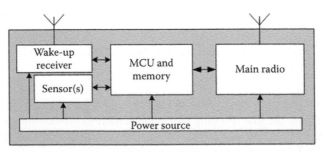

FIGURE 5.4   Sensor node architecture for dual-radio approach.

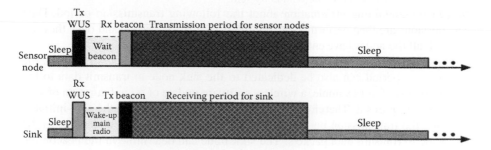

FIGURE 5.5   Source-initiated mode of the GWR-MAC protocol.

In the source-initiated mode, the sensor node(s) will wake up the sink node from the sleep mode by transmitting a WUS, as illustrated in Figure 5.5. To decrease the probability of WUS collisions, a random (or predefined) delay for WUS transmissions can be used. The WUR of the sink node will receive the WUS and generate a wake-up via microcontroller (MCU) to the main radio. The sink node's main radio will then broadcast a beacon (BC) message to initiate transmission period for the sensor node(s) according to the channel access procedure, which can be based on different methods suitable for different scenarios. The beacon message is, at the same time, an acknowledgement to the sensor node(s) that the WUS has been received by the sink. If the sensor node does not get the beacon message, it will retransmit the WUS after a random back-off period. The WUS transmission procedure is therefore similar to the Aloha channel access with a random (or predefined) delay for the first transmission. Once the beacon message is received, the sensor node(s) will send the data packet(s) to the sink during the transmission period by using the channel access method informed in the beacon message. This mode of the GWR-MAC protocol is therefore a combination of the source-initiated wake-up procedure and the following channel access control method for the transmission period.

In the sink-initiated mode, illustrated in Figure 5.6, the sink node will wake up the sensor nodes from the sleep mode by sending the WUS using broadcast, unicast, or multicast. It depends on the used WUR technology whether it is possible to use addressing to wake up only certain sensor nodes or whether broadcast should be used. When the sensor node receives the WUS, it will send an ACK message to the sink. The sink node knows that the WUS has been detected correctly and sends

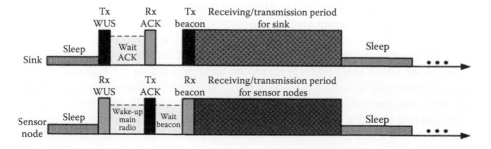

FIGURE 5.6   Sink-initiated mode of the GWR-MAC protocol.

the beacon containing information about the following transmission period. Data transmissions are then performed during the transmission period, and once they are finished, all the nodes have entered the sleep mode.

Typically, the data flow is from the sensor nodes to the sink node. However, the transmission period can also be dedicated to the sink node to transmit data to the sensor node(s), if, for example, a wireless software update or reconfiguration of sensor node(s) requires it. Therefore, in Figure 5.6, it is illustrated for the sink-initiated case that the transmission period can be dedicated to the sensor node(s) or to the sink node to transmit data packets. The sink node can determine in the beacon the upcoming transmission period channel access mechanism, timing, and scheduling information. Different channel access methods can be used for the transmission period management. When the transmission period is finished, all the nodes have entered the sleep mode and the next transmission period will take place after the next wake-up procedure.

For the transmission period channel access, one option is to use a contention-based MAC. In that case, the nodes compete for channel access and transmit packets according to the contention-based MAC principle. For example, in the Aloha case, nodes transmit when they have a packet to transmit. If the packet is not successfully received, then the retransmission policy defines either that the packet is discharged or retransmitted. For example, in the Aloha case, the unsuccessful packet will be retransmitted again after a random back-off period if the ACK is not received during a certain time. In addition, other contention-based methods, for example, carrier sensing multiple access with collision avoidance, and so on, can be used during the transmission period. Another option is to use contention-free scheduled methods, for example, time division multiple access–based protocols, guaranteed time slots defined in the IEEE Std 802.15.4 [37], and scheduled access mode of the IEEE Std 802.15.6 [38] or ETSI SmartBAN MAC [39]. The requirement for the usage of contention-free methods is that the sink node assigns dedicated time slots for each sensor node. The sink node does not have information about which nodes have a packet to transmit, and therefore the channel resources may be wasted. In the ideal contention-free case, collisions will not occur if the nodes are perfectly synchronized and follow the schedule.

The described GWR-MAC protocol principle is suitable for different types of application since it defines a bidirectional wake-up procedure between the sensor nodes and the sink. In addition, it enables the usage of different channel access methods for the transmission period. However, note that the GWR-MAC protocol is not defined here in a detailed manner (wake-up collision avoidance, packet formats, turnaround times, etc.). Instead, the described GWR-MAC is a general-level framework for short-range networks that take advantage of the wake-up radios. Which mode of the GWR-MAC protocol should be used depends on the application scenario. For example, in some applications, only the source-initiated case may be used, and then only the sink node should be equipped with a WUR. In some scenarios, there can be a need that only the sink node must be able to wake up sensor nodes. In that case, the sink-initiated mode would be used and only the sensor nodes must be accompanied with a WUR. Some application scenarios require both modes of the

GWR-MAC protocol. In such cases, all the network nodes must be equipped with a WUR and data radio as was illustrated in Figure 5.4.

5.3   WUR SOLUTIONS

WUR designs have progressed substantially in the last decade and they will most probably be utilized in various commercial medical and IoT applications in the near future. The current designs consume power well below 100 μW, which is already more than hundred times less than what a typical commercial RF transceiver designed for WSNs consumes. Low consumption enables a node to listen to the channel constantly for years with a single coin size battery, but because of the trade-off between power consumption and sensitivity, sensitivities of WURs are usually considerably worse compared to commercial RF transceivers. Consequently, to reach the WUR and the RF transceiver of the node, more transmit power is needed to transfer a WUS to the WUR than to transmit data to the RF transceiver.

Receiver design always includes trade-offs regarding, for example, selectivity, data rate, power consumption, and sensitivity. The two main performance comparison metrics of the proposed solutions have been power consumption and sensitivity, followed by data rate and selectivity. In some works, energy per bit is also used as a comparison metric. However, utilized wake-up packet lengths are usually only a few bytes in size; hence, the receiver's active power consumption becomes more important than energy per bit. Also, because of the small size of the wake-up packet, it is transmitted reasonably fast even with low data rates. The different receiver architectures that can be used for WUS detection will be discussed in more detail in Section 5.3.1. Furthermore, future research challenges will be outlined in Section 5.3.2.

5.3.1   WUR ARCHITECTURES

Data receivers typically use direct conversion architecture, but because of the challenges of flicker noise and DC offset, they are rarely used in WURs. Also, the common superheterodyne receiver architecture has been proposed to be used as a WUR. However, in those solutions, duty cycling is often employed and active power consumptions of different superheterodyne solutions are significantly more than those in WURs on the basis of other architectures. Because of the demanding requirements of power consumption and sensitivity, receiver designers have proposed many different receiver architectures for WUR purposes. Figure 5.7 shows block diagrams of the most common architectures, which are based on RF envelope detection (RFED), uncertain IF, matched filter, injection locking, superregenerative oscillator, and subsampling.

Being the most straightforward option for energy detection, the RFED architecture is the most commonly used solution. In RFED-based receivers, a local oscillator is not needed since the envelope of the RF signal is detected and signal is therefore directly down converted to baseband. The envelope is detected from a wide bandwidth. Therefore, the incoming signal is band pass filtered in order to ensure low noise at the envelope detector input.

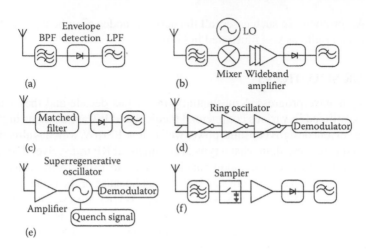

FIGURE 5.7   Typical WUR architectures: (a) RF envelope detection, (b) uncertain IF, (c) matched filter, (d) injection locking, (e) superregenerative oscillator, and (f) subsampling.

Accurate RF synthesizers typically consume too much power to be used in WURs. The uncertain-IF receiver architecture utilizes a synthesizer that consumes low power at the cost of poor frequency stability. After the signal is down converted, the signal is located in a wide frequency range. Wideband amplification is typically used to improve SNR followed by envelope detection.

Majority of WURs use narrowband WUSs even though in-band interference is a major challenge for receiver architectures that are based on received signal power. With a spread spectrum technique, the signal is spread in the frequency domain, which makes it more resistant against interference owing to spreading gain that will be achieved by dispreading the signal at the receiver. To dispread the spread spectrum signal energy efficiently, for example, passive surface acoustic wave (SAW) matched filter can be used. Low-power consumption is achieved at the cost of sensitivity because of high insertion losses of the SAW matched filter.

A receiver based on injection-locking architecture has an oscillator that locks to a received carrier frequency if it is close to the oscillator's natural oscillation frequency. If injection locking does not occur, but the carrier frequency disturbs the oscillator, it will be seen at the oscillator output. This is called injection polling. Hence, injection-locking receivers are usually used to detect frequency shift keying–modulated signals.

The superregenerative receiver is based on power detection. Oscillation start-up time in a superregenerative oscillator correlates with the received signal strength. By detecting time difference between oscillation start-up times, a simple low-power receiver that has large gain can be built. The superregenerative oscillator architecture block diagram in Figure 5.7e has a low-noise amplifier (LNA) at the front-end, because it is needed to isolate feed-through from the oscillator to the antenna and it also amplifies the signal. It depends on superregenerative receiver characteristics that the oscillator emits to the surroundings. If emission is larger than regulations allow, isolation must be implemented. LNA can be added at the front-end of other architectures as well, which would give gain at the cost of power consumption.

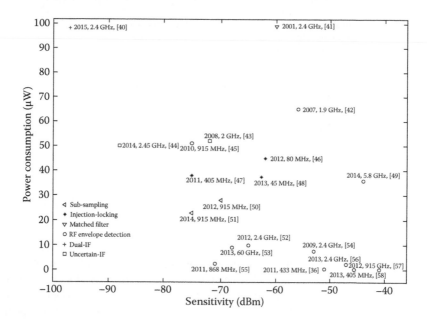

FIGURE 5.8 Comparison of WURs (Refs. [36], [40–58]) and their architectures.

Typically in WURs, amplification is rather done at IF than in RF because it is more power friendly.

Subsampling-based receiver architecture down converts the signal with a discrete-time sampler instead of a continuous-time mixer. Input signal is mixed with harmonic components in the discrete-time sampler, producing replicas in multiple frequencies that give more freedom to select the appropriate IF.

Different WUR designs are compared in Figure 5.8, where the trade-off between sensitivity and power consumption is clearly illustrated. At the moment, there is no superior architecture that would be clearly more suitable for WUR purposes in comparison to other solutions.

5.3.2 FUTURE CHALLENGES

WUR sensitivities vary from −40 dBm to around −90 dBm in designs that have active power consumption less than 50 μW, whereas typical commercial short-range transceivers usually have a sensitivity of at least −95 dBm. Therefore, there is still work to be done to reach the gap between WURs and data transceivers. If the wake-up range is lower than the data radio range, the usage of wake-up signaling will require a higher node density in the network or a higher transmit power. The gap becomes even a larger issue if one would equip WURs to wide-area IoT sensor nodes. In a wide-area IoT network (also called as low power wide area network [LPWAN]), the link distance can be more than 10 km, whereas in a traditional local-area sensor network, distances are usually some tens of meters. Figure 5.9 shows the main differences between a local-area and a wide-area IoT sensor network. Medical applications for wide-area IoT sensor networks could be, for example, remote patient monitoring.

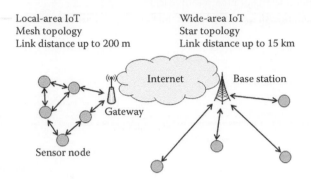

Local-area IoT
Mesh topology
Link distance up to 200 m

Wide-area IoT
Star topology
Link distance up to 15 km

FIGURE 5.9　Differences between local-area and wide-area IoT sensor networks.

Sensor nodes can be configured to periodically report a patient's well-being directly via a remote base station, which enables the patient to not carry the gateway device. In normal conditions, sensor nodes can be configured to transmit a patient's vital data to the base station a couple of times per day, but if the patient's physical condition has some alarming signs, and the doctor needs data more frequently, WUR usage enables the reporting frequency to be changed with low latency. It is expected that this kind of situation occurs rarely; therefore, it is assumed to be more energy efficient to use nodes that are equipped with WUR instead of duty cycling. In order to achieve a 10-km range with a transmit power that regulations allow, commercial wide-area IoT transceivers are operating in sub–1-GHz bands and they are designed to be highly sensitive. For example, the SX1272 transceiver by Semtech [59] achieves a sensitivity of −137 dBm at a data rate of 300 bps. Semtech is part of the LoRa alliance, which is targeting to standardize a solution for the wide-area IoT field. Another technology provider, focused on wide-area IoT, called SigFox uses transceivers such as AX5043 by Axsem [60], which has a similar performance to SX1272. Most of the WURs found from the literature have much higher data rates, from 100 to 350 kbps. One direction for future WUR designs could be therefore sensitivity improvement at the cost of data rate. Since the length of the wake-up packet is usually in the order of a few bytes, the WUS can be transmitted sufficiently fast even with low data rates while maintaining energy efficiency.

The world is rapidly evolving into a networking society where more and more wireless devices communicate with each other. Gartner, a technology research and advisory firm, estimated in 2014 that there will be 26 billion IoT-based devices in 2020 [61]. ABI Research, a technology market intelligence firm, estimated the number of devices to be 30 billion [62], and Cisco's estimate is that 50 billion devices and objects will be connected to IoT by 2020 [8]. These estimates show that the IoT market is potentially huge. Assuming that most of the devices will be wireless, the number of radios can be even higher since a device might be equipped with multiple radios. More wireless devices mean that there might be more interference, but on the other hand, new millimeter wave bands will be exploited that will ease the situation. More interference will be harmful, especially for the RF envelope detection architecture since it is based on energy detection. Perhaps more interference-robust WUR architectures will be more useful in the near future.

WUS generation and transmission are often neglected in the WUR design and energy efficiency performance evaluation. However, since OOK is the most common modulation method being used, it can be generated with most of the commercial transceivers. Also, data rate can be usually configured to match the receiver's rate. In the wide-area IoT case, the base stations are already being installed all around the world by SigFox and LoRa alliances; therefore, the hardware changes would be expensive. In order to make WURs a widespread technology in the future, integrating WURs to existing and future IoT networks should get more attention in the research.

5.4 ENERGY-EFFICIENT TARGET SCENARIOS

5.4.1 APPLICATION SCENARIOS

The hierarchical architecture is scalable for various application scenarios that are deployed using heterogeneous devices. However, the wake-up radio–based hierarchical architecture is targeted especially to applications that require communication rarely and in addition require a low latency reaction to events.

A very good example is an area surveillance network. In that case, the motion detection sensor (e.g., passive infrared) nodes are continuously monitoring the environment to detect intruder movements in the sensing area. Once the event is detected, sensor nodes must be able to report the event rapidly. In the energy-efficient hierarchical architecture case, it means that the next layer in the hierarchy must be awakened with a very low delay. If the event is detected to be critical, the highest layer will also be awakened and an alarm will be generated to the user through the backbone. Another example is structural monitoring of, for example, bridges or buildings. In that case, the condition of the monitored structure can be queried rarely. On the other hand, the sensor nodes can send an alarm if some critical changes have been detected. Also, in that case, the sensor nodes can perform continuous monitoring and wake up the higher-layer node only when required in order to enable long lifetime for the network nodes. Hierarchical wake-up radio–based architecture can be as well utilized for WBAN and industrial applications. In the WBAN case, the wake-up radios could be equipped for instance to implants and on-body nodes that must have very long lifetimes once installed in the human body. In industrial applications, WUR-based nodes can be installed to monitor, for example, pipe valves or certain parts of engines. For military and critical infrastructure scenarios, there are also many uses for hierarchical WUR-based network architecture since there is a need for different types of long-term monitoring applications, for example, for surveillance, reconnaissance, and security purposes. Figure 5.10 summarizes example applications and communication and sensor technologies that can be used in their implementation.

5.4.2 ENERGY EFFICIENCY COMPARISON OF WAKE-UP MECHANISMS

This section introduces an energy efficiency comparison of the WUR- and DCR-based wake-up mechanisms discussed in Section 5.2.2. The intelligent hierarchical operation is assumed for both WUR and DCR approaches; that is, the lower-layer

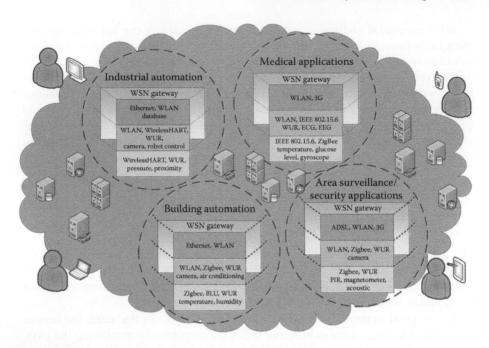

FIGURE 5.10 Examples of hierarchical WSN architecture application areas and techniques.

devices of the hierarchical architecture are performing continuous sensing and will wake up the higher layers when required. At first, the comparison is done with different duty cycle and event frequency values in a typical surveillance scenario for critical infrastructure protection, which corresponds with the case of subnetwork 2 in Figure 5.2. For both approaches, source-initiated communication is assumed; that is, the focus is on the scenario where the sensor nodes have sensed something and trigger the wake-up of the higher layer. The used analytical model can also be applied to other WSN scenarios with layered hierarchical architecture, but in that case, the detailed assumptions must be adjusted to match the particular application under evaluation.

In the DCR approach, each layer's low-power transceivers need to wake up and sleep according to a predefined schedule. It is assumed that the nodes in each network layer have low-power radios (e.g., IEEE Std 802.15.4-based) for duty cycling to enable the wake-up mechanism. In the DCR-based network case, the transmitter sends a DE message in order to inform the node at a higher layer that it should stay awake. If a higher-layer node receives a DE message during the duty cycle listening, it will stay on, send an ACK message, and wait for a data transmission.

In the WUR-based network, the WUSs are used to activate the higher layers when needed. The GWR-MAC protocol introduced in Section 5.2.3 is used in the WUR-based networks. The lower-layer node transmits the WUS and the higher-layer node sends back a beacon message, as described in the GWR-MAC protocol's source-initiated mode definition. Then, the lower-layer nodes can send their data to the higher-layer node. The same GWR-MAC wake-up procedure is used for the EL to wake up the IL and the IL to wake up the AL.

We have originally proposed an analytical model, which can be used to compare the energy efficiency of the GWR-MAC–based and conventional duty cycle MAC (DCM)–based networks as a function of number of events in a hierarchical network for surveillance scenario [22]. The energy efficiency comparison takes account of the energy consumption of the nodes' core components: MCU unit, transceiver, and sensors. A low-power MCU has typically the active, standby, and sleep modes. The transceiver has the transmit (Tx), receive (Rx), idle, and sleep modes. The active, Tx, and Rx modes are the most energy-consuming parts. Therefore, it is important to put the nodes into the sleep mode when possible. The sensing component consists of sensors and A/D converters. The sensing component has different modes, which affect energy consumption, for example, sensor warm-up, active mode, and settle time of the A/D converter. The dominating energy consumption factors of each transceiver's components are taken here into account: wake-up signaling, data transmission and reception, and MCU and sensor active mode current consumption. The relevant energy consumption characteristics, affecting the WUR and DCR energy efficiency comparison, are then addressed. However, energy consumption during network initialization and run-time management (e.g., communication required for synchronization and routing) are not taken into account.

The energy efficiency equation for the WUR and DCR network comparison is defined as [22,32]

$$\eta(\varepsilon,\lambda,t,\beta) = \frac{\min(E(\varepsilon,\Lambda,t,\beta))}{E(\varepsilon,\lambda,t,\beta)}, \qquad (5.1)$$

where E is the network energy consumption over time period t, ε is the number of events during t, λ is duty cycle, and β is bit error probability. In this case, the event means that nodes have a data packet to send for a hub about the sensed event. In Equation 5.1, the minimum of E is calculated over the duty cycle value set $\Lambda = (0,1]$. Note that $\Lambda = 1$ corresponds to the WUR case since the receiver is listening to the channel continuously. The metric introduced in Equation 5.1 defines the maximum energy efficiency to be one and enables comparison of the WUR- and DCR-based networks.

The total energy consumption during the operation time, t, as a function of number of events and bit error probability, for WUR-based network layers (EL, IL, and AL), can be calculated as

$$E_{\text{WUR}}^{\text{EL}}(\varepsilon,t,\beta) = E_{\text{s}}^{\text{EL}}(t) + E_{\text{MCU}}^{\text{EL}}(\varepsilon,t) + E_{\text{TX,WUS}}(\varepsilon,t,\beta) + E_{\text{wait,BC}}(\varepsilon,t) + E_{\text{RX,BC}}(\varepsilon,t)$$

$$+ E_{\text{C}}(t) + E_{\text{TX,D}}^{\text{EL}}(\varepsilon,t,\beta) + E_{\text{clk}}(t)$$

$$E_{\text{WUR}}^{\text{IL}}(\varepsilon,t,\beta) = E_{\text{s}}^{\text{IL}}(t) + E_{\text{MCU}}^{\text{IL}}(\varepsilon,t) + E_{\text{TX,WUS}}(\varepsilon,t,\beta) + E_{\text{TX,BC}}(\varepsilon,t) + E_{\text{wait,BC}}(\varepsilon,t)$$

$$+ E_{\text{RX,BC}}(\varepsilon,t) + E_{\text{RX,WUS}}(\varepsilon,t,\beta) + E_{\text{C}}(t) + E_{\text{clk}}(t) + E_{\text{TX,D}}^{\text{IL}}(\varepsilon,t,\beta) + E_{\text{RX,D}}^{\text{IL}}(\varepsilon,t,\beta)$$

$$E_{\text{WUR}}^{\text{AL}}(\varepsilon,t,\beta) = E_{\text{MCU}}^{\text{AL}}(\varepsilon,t) + E_{\text{RX,WUS}}(\varepsilon,t,\beta) + E_{\text{TX,BC}}(\varepsilon,t) + E_{\text{C}}(t) + E_{\text{clk}}(t)$$

$$+ E_{\text{TX,D}}^{\text{AL}}(\varepsilon,t,\beta) + E_{\text{RX,D}}^{\text{AL}}(\varepsilon,t,\beta),$$

$$(5.2)$$

where $E_{\text{TX,WUS}}$ is the energy consumption of WUS transmissions, $E_{\text{RX,WUS}}$ is the energy consumption of WUS receptions, $E_{\text{TX,BC}}$ is the energy consumption of beacon transmission, $E_{\text{wait,BC}}$ is the energy consumption of beacon listening, $E_{\text{RX,BC}}$ is the energy consumption of beacon receptions, E_{C} is the constant energy consumption of WUR, and E_{clk} is the energy consumption of the clock needed to maintain the time synchronization. E_{s}^{x} is the energy consumption of sensing, E_{MCU}^{x} is the energy consumption of MCU, and $E_{\text{TX,D}}^{x}$ and $E_{\text{RX,D}}^{x}$ are the energy consumption of data transmissions and receptions, respectively, calculated separately for each layer (i.e., x is EL, IL, or AL).

The total energy consumption during t, as a function of number of events, duty cycle percentage, and bit error probability, for DCR-based network layers (EL, IL, and AL) can be calculated as

$$E_{\text{DCR}}^{\text{EL}}(\varepsilon,\lambda,t,\beta) = E_{\text{s}}^{\text{EL}}(t) + E_{\text{MCU}}^{\text{EL}}(\varepsilon,t) + E_{\text{RX,DC}}^{\text{EL}}(\lambda,t) + E_{\text{clk}}(t) + E_{\text{TX,DE}}^{\text{EL}}(\varepsilon,t,\beta)$$

$$+ E_{\text{RX,BC}}^{\text{EL}}(\varepsilon,t,\beta) + E_{\text{TX,D}}^{\text{EL}}(\varepsilon,t,\beta)$$

$$E_{\text{DCR}}^{\text{IL}}(\varepsilon,\lambda,t,\beta) = E_{\text{s}}^{\text{IL}}(t) + E_{\text{MCU}}^{\text{IL}}(\varepsilon,t) + E_{\text{RX,DC}}^{\text{IL}}(\lambda,t) + E_{\text{clk}}(t) + E_{\text{RX,DE}}^{\text{IL}}(\varepsilon,t,\beta)$$

$$+ E_{\text{TX,BC}}^{\text{IL}}(\varepsilon,t,\beta) + E_{\text{TX,DE}}^{\text{IL}}(\varepsilon,t,\beta) + E_{\text{RX,BC}}^{\text{IL}}(\varepsilon,t,\beta) + E_{\text{TX,D}}^{\text{IL}}(\varepsilon,t,\beta) + E_{\text{RX,D}}^{\text{IL}}(\varepsilon,t,\beta)$$

$$E_{\text{DCR}}^{\text{AL}}(\varepsilon,\lambda,t,\beta) = E_{\text{MCU}}^{\text{AL}}(\varepsilon,t) + E_{\text{RX,DC}}^{\text{AL}}(\lambda,t) + E_{\text{clk}}(t) + E_{\text{RX,DE}}^{\text{AL}}(\varepsilon,t,\beta)$$

$$+ E_{\text{TX,BC}}^{\text{AL}}(\varepsilon,t,\beta) + E_{\text{TX,D}}^{\text{AL}}(\varepsilon,t,\beta) + E_{\text{RX,D}}^{\text{AL}}(\varepsilon,t,\beta),$$

$$(5.3)$$

where λ is the duty cycle percentage, $E_{\text{RX,DC}}^{x}$ is the energy consumption of channel listening according to the duty cycle, and $E_{\text{TX,DE}}^{x}$ and $E_{\text{RX,DE}}^{x}$ are the energy consumption of DE message transmission and reception, respectively, when x is EL, IL, or AL. Depending on the duty cycling–based MAC protocol features, the DE message can be replaced, for example, by using a preamble before the data packet.

By multiplying the energy consumption of different layer nodes at a WUR-based network (Equation 5.2) and a DCR-based network (Equation 5.3) with the number of nodes at each layer, the network total energy consumption during operation time, t, can be easily derived. Interested readers can find more details about the network energy consumption derivation from Refs. [22,32].

Figure 5.11 shows network total energy consumption comparison results, which are calculated using the parameters chosen to represent typical values for nodes equipped with WUR, IEEE 802.15.4, and IEEE 802.11b communication interfaces and sensors, as described for the surveillance scenario. The number of nodes at the EL is 100, and the number of nodes at the IL is 10; that is, there is on average 10 sensor nodes assumed to be associated with one IL node, which acts as a coordinator node for the sensor nodes. It can be seen that in the WUR-based network, energy consumption is drastically lower with the whole range of studied duty cycle values when the event frequency is low. For the lowest number of events case, the energy consumption gain of the WUR approach is more than two orders of magnitude in

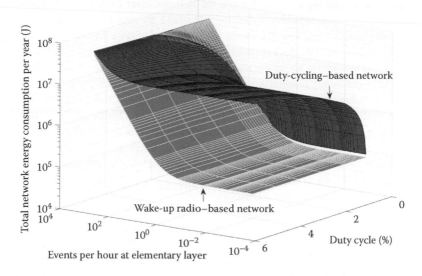

FIGURE 5.11 Network energy consumption comparison as a function of event per hour and duty cycle.

comparison with a DCR with $\lambda = 5\%$. For the highest number of events per hour, DCR has an energy consumption gain of 12% with the smallest duty cycle percentage value. More details about the analytical model, parameters, and results can be found in Ref. [22].

In Ref. [31], the authors have shown that hierarchical architecture and its analytical energy efficiency model introduced in Ref. [22] can also be applied for the WBAN scenario. Further, in Ref. [28], the wake-up radio– and DCR-based network energy consumption comparison is discussed in WBAN's case. Here, WUR and DCR energy efficiency comparison results are shown, which are calculated using four different WUR parameters, as shown in Table 5.1. Typical state-of-the-art performance values are used in WUR1 and WUR2 for power consumption and sensitivity. WUR3

TABLE 5.1

Parameters Used for Different Radios in the Energy Efficiency Comparison

				Power Consumption	
Radio	Sensitivity	Tx Power	Data Rate	Tx Mode [63]	Rx Mode
WUR1	−70 dBm	0 dBm	200 kbps	52.2 mW	5 μW
WUR2	−80 dBm	−10 dBm	200 kbps	33.9 mW	10 μW
WUR3	−95 dBm	−25 dBm	200 kbps	25.5 mW	50 μW
WUR4	−95 dBm	−25 dBm	300 bps	25.5 mW	5 μW
DCR	−95 dBm	−25 dBm	971 kbps	25.5 mW	56 mW [63]

sensitivity is set to be the same as for the DCR transceiver [63]. In the WUR4 case, the assumption is that sensitivity can be improved by using a very low data rate while maintaining a very low Rx mode power consumption. In the calculations, it was assumed that the data packet payload is 255 bytes and communication is error free ($\beta = 0$). More details about the used analytical model can be found in Ref. [31].

The energy efficiency comparison results for WUR- and DCR-based WBANs as a function of number of events per hour is presented in Figure 5.12. It can be observed that the GWR-MAC–based network outperforms the DCM-based network's lowest duty cycle ($\lambda = 0.5\%$) case when the number of events is less than 12 per hour. When compared to the DCM network with $\lambda = 3\%$, the GWR-MAC–based approach is more energy efficient if the number of events is below 60 per hour. The results for different WURs show that WUR's Rx mode power consumption has a remarkable effect on the total energy efficiency since it is continuously listening to the channel to detect WUSs. The WUR1-based network features the lowest sensitivity for WUR. However, it is the most energy efficient when $\varepsilon < 12$, even though it requires the highest power used by the transmitter. The WUR3 case has the highest Rx mode power consumption, which leads to drastically lower energy efficiency if events occur rarely. This observation highlights the importance of constant mode power consumption (Rx mode) minimization for WURs. All the studied WUR cases lead to higher energy efficiency than the DCR approach when the number of events is less than 12 per hour. When the number of events increases above 12 per hour, the energy consumption of data communication starts to dominate in the network overall energy consumption and the difference between WUR's energy efficiency is not visible anymore. Furthermore, it can be observed that if the sensitivity of WUR can be

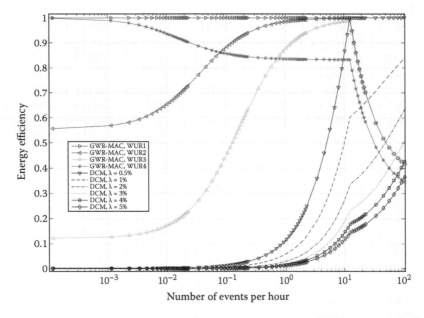

FIGURE 5.12 Energy efficiency comparison for WUR-based and DCM-based hierarchical network.

improved by decreasing the data rate, it will lead to an energy-efficient solution for very rare event cases. When the event frequency increases, the longer transmission and reception duration will cause lower energy efficiency in comparison to higher data rate WUR solutions. From the results of Figure 5.12, it can be concluded that the WUR-based approach is drastically more energy efficient than the duty cycle–based approach when the event frequency is low.

5.5 SUMMARY

Hierarchical architecture for WSNs that can be used to provide data and services to medical applications and IoT was discussed in this chapter. The introduced architecture is designed to enable the deployment of multiple different technologies in the same network. Therefore, it can be used for different types of monitoring scenarios, for example, sensor networks for medical patients and monitoring of patients, as well as for many other types of WSN and WBAN applications. Functionalities for different layers are designed so that the network can fulfill its requirements with low power operation. Energy efficiency is achieved by utilizing wake-up signaling that can be used to activate the layers only when required. A GWR-MAC protocol designed for hierarchical architecture was introduced. The GWR-MAC protocol includes a bidirectional wake-up procedure and data transmission period in which the channel access method can be selected depending on the application characteristics. Different WUR architectures were introduced and future challenges were outlined. Architecture energy efficiency performance results were shown for a typical area surveillance scenario and WBAN case. Results show that WUR-based networks have a remarkable potential to improve energy efficiency, in comparison to traditional duty cycle operation, in the case of low event rate applications. The DCR approach is more energy efficient only when sufficiently low duty cycle is combined with a high number of events. However, in many practical solutions, the duty cycle is fixed and must be large enough to handle the worst-case traffic. Duty cycle should be changed dynamically when the event frequency changes in order to save energy. Furthermore, the very low duty cycle operation would require very strict synchronization in order for transmissions to be done exactly at correct times according to the duty cycle. Strict synchronization maintenance requires message exchange between the network nodes, which will cause additional energy consumption. The proposed GWR-MAC protocol needs to be defined in more detail and implemented to verify the functionalities and performance in real applications. The wake-up radio research area is still quite unexplored, and therefore the purpose of the hierarchical architecture with the GWR-MAC protocol is to enable more efficient usage of WURs in WSNs and WBANs and to foster future research and development.

REFERENCES

1. Raghunathan V, Schurgers C, Park S and Srivastava M. 2002. Energy-aware wireless microsensor networks. *IEEE Signal Processing Magazine* 19(2): 40–50.
2. Pottie GJ and Kaiser WJ. 2000. Wireless integrated network sensors. *Communications of the ACM* 43(5): 51–58. Available at http://doi.acm.org/10.1145/332833.332838.

3. Ares BZ, Park PG, Fischione C, Speranzon A and Johansson KH. 2007. On power control for wireless sensor networks: System model, middleware component and experimental evaluation. In: *Proc. European Control Conference*, pp. 1–8.
4. Hohlt B, Doherty L and Brewer E. 2004. Flexible power scheduling for sensor networks. In: *Proc. Third International Symposium on Information Processing in Sensor Networks*. IPSN 2004, pp. 205–214.
5. Demirkol I, Ersoy C and Alagoz F. 2006. MAC protocols for wireless sensor networks: A survey. *IEEE Communications Magazine* 44(4): 115–121.
6. Anastasi G, Conti M, Francesco MD and Passarella A. 2009 Energy conservation in wireless sensor networks: A survey. *Ad Hoc Networks* 7(3): 537–568.
7. Nechibvute A, Chawanda A, and Luhanga P. 2012. Piezoelectric energy harvesting devices: An alternative energy source for wireless sensors. *Smart Materials Research*, 2012, Article ID 853481, pp. 1–13. doi: 10.1155/2012/853481.
8. Evans D. 2011. The internet of things: How the next evolution of the internet is changing everything. Cisco Internet Business Solutions Group (IBSG) white paper, April.
9. Buratti C and Verdone R. 2008. A hybrid hierarchical architecture: From a wireless sensor network to the fixed infrastructure. In: *Proc. Wireless Conference*, pp. 1–7, EW 2008.
10. Khan Z, Catalot D and Thiriet J. 2009. Hierarchical wireless network architecture for distributed applications. In: *Proc. Wireless and Mobile Communications*, ICWMC '09, pp. 70–75.
11. Yang J, Gao Y and Zhang Z. 2011. Cluster-based routing protocols in wireless sensor networks: A survey. In: *Proc. International Conference on Computer Science and Network Technology*, vol. 3 of ICCSNT '11, pp. 1659–1663.
12. ZigBee Alliance. 2014. ZigBee Alliance webpage. Available at http://www.zigbee.org.
13. Z-Wave Alliance. 2014. Z-Wave alliance webpage. Available at http://www.z-wavealliance .org.
14. Kumar D, Aserib TC and Patelc RB. 2009. EEHC: Energy efficient heterogeneous clustered scheme for wireless sensor networks. *Journal of Computer Communications* vol. 32(4): pp. 662–667.
15. Wu M and Collier M. 2011. Extending the lifetime of heterogeneous sensor networks using a two-level topology. In: *Proc. IEEE International Conference on Computer and Information Technology*, pp. 499–504.
16. Slimane JB, Song YQ, Koubâa A and Frikha M. 2009. A three-tiered architecture for large scale wireless hospital sensor networks. In: *Proc. The International Workshop on Mobilizing Health Information to Support Healthcare-related Knowledge Work*, MobiHealthInf 2009, pp. 1–12.
17. Zhang M, Song J and Zhang Y. 2005. Three-tiered sensor networks architecture for traffic information monitoring and processing. In: *Proc. Intelligent Robots and Systems*, IROS 2005, pp. 2291–2296.
18. Kulkarni P, Ganesan D, Shenoy P and Lu Q. 2005. SensEye: A multitier camera sensor network. In: *Proc. ACM International Conference on Multimedia*, MM '05, pp. 229–238.
19. Lopes CER, Linhares FD, Santos MM and Ruiz LB. 2007. A multi-tier, multimodal wireless sensor network for environmental monitoring. *Springer Link Lecture Notes in Computer Science* 4611: 589–598.
20. Zatout Y, Campo E and Llibre JF. 2009. WSN-HM: Energy-efficient wireless sensor network for home monitoring. In: *Proc. International Conference on Intelligent Sensors, Sensor Networks and Information Processing*, ISSNIP '09, pp. 367–372.
21. Stefanov A and Stojanovic M. 2010. Hierarchical underwater acoustic sensor networks. In: *Proc. ACM International Workshop on UnderWater Networks*, WUWNet'10, pp. 1–4.

22. Karvonen H, Suhonen J, Petäjäjärvi J, Hämäläinen M, Hännikäinen M and Pouttu A. 2014. Hierarchical architecture for multi-technology wireless sensor networks for critical infrastructure protection. *Springer Wireless Personal Communications Special Issue on Intelligent Infrastructures*, 76(2): 209–229. doi: http://dx.doi.org/10.1007/s11277 -014-1686-2.

23. Crossbow. 2014. TelosB datasheet. Available at http://www.willow.co.uk/TelosB _Datasheet.pdf.

24. FriendlyARM. 2014. FriendlyARM Mini6410 specification. Available at http://www .friendlyarm.net/products/mini6410.

25. Kuorilehto M, Kohvakka M, Suhonen J, Hämäläinen P, Hännikäinen M and Hämäläinen T. 2007. *Ultra-Low Energy Wireless Sensor Networks in Practice: Theory, Realization and Deployment*. Chichester, West Sussex, England: John Wiley & Sons.

26. Le TN, Magno M, Pegatoquet A, Berder O, Sentieys O and Popovici E. 2013. Ultra low power asynchronous MAC protocol using wake-up radio for energy neutral WSN. In: *Proc. International Workshop on Energy Neutral Sensing Systems*, ENSSys '13, pp. 10:1–10:6. ACM, New York. Available at http://doi.acm.org/10.1145 /2534208.2534221.

27. Mahlknecht S and Spinola Durante M. 2009. WUR-MAC: Energy efficient wakeup receiver based MAC protocol. In: *Proc. IFAC International Conference on Fieldbuses & Networks in Industrial & Embedded Systems*, FET 2009, pp. 139–143.

28. Petäjäjärvi J, Karvonen H, Vuohtoniemi R, Hämäläinen M and Huttunen M. 2014. Preliminary study of superregenerative wake-up receiver for WBANs. In: *Proc. International Symposium on Medical Information and Communication Technology* (ISMICT'14), pp. 1–5.

29. Petäjäjärvi J, Karvonen H, Mikhaylov K, Pärssinen A, Hämäläinen M and Iinatti J. 2015. WBAN energy efficiency and dependability improvement utilizing wake-up receiver. *IEICE Transactions on Communications—Special Issue on Innovation of Medical Information and Communication Technology for Dependable Society*, E98-B (04): 535–542, Apr.

30. Karvonen H, Petäjäjärvi J, Iinatti J, Hämäläinen M and Pomalaza-Ráez C. 2014. A Generic wake-up radio based MAC protocol for energy efficient short range communication. *IEEE PIMRC Workshop: The Convergence of Wireless Technologies for Personalized Healthcare*, Sept. 2–5, Washington, DC, USA.

31. Karvonen H, Petäjäjärvi J, Iinatti J and Hämäläinen M. 2014. Energy Efficient IR-UWB WBAN using a Generic Wake-up Radio based MAC Protocol. The Third Ultra Wideband for Body Area Networking Workshop (UWBAN-2014), co-located with the 9th International Conference on Body Area Networks (BodyNets-2014), Sept. 29–Oct. 1, London, UK.

32. Karvonen H. 2015. *Energy Efficiency Improvements for Wireless Sensor Networks by Using Cross-Layer Analysis*. PhD thesis, University of Oulu, Finland, March.

33. Gu L and Stankovic JA. 2005. Radio-triggered wake-up for wireless sensor networks. *Springer Journal on Real-Time Systems* 29(2): 157–182.

34. Van der Doorn B, Kavelaars W and Langendoen K. 2009. A prototype low-cost wakeup radio for the 868 MHz band. *International Journal of Sensor Networks* 5(1): 22–32.

35. Ansari J, Pankin D and Mähönen P. 2009. Radio-triggered wake-ups with addressing capabilities for extremely low power sensor network applications. *International Journal of Wireless Information Networks* 16(1): 118–130.

36. Marinkovic SJ and Popovici EM. 2011. Nano-power wireless wake-up receiver with serial peripheral interface. *IEEE Journal on Selected Areas in Communications* 29(8): 1641–1647.

37. IEEE. 2011. IEEE Standard for Local and Metropolitan Area Networks—Part 15.4: Wireless Medium Access Control (MAC) and Physical Layer (PHY) Specifications for Low-Rate Wireless Personal Area Networks (LR-WPANs). Standard, The Institute of Electrical and Electronics Engineers, Inc. IEEE Std. 802.15.4-2011, Revision of IEEE Std 802.15.4-2006.

38. IEEE Std. 802.15.6. 2012. IEEE Standard for Local and Metropolitan Area Networks—Part 15.6: Wireless Body Area Networks. The Institute of Electrical and Electronics Engineers, Inc, Standard.

39. ETSI TC SmartBAN. 2014. Smart Body Area Networks (SmartBAN); Low Complexity Medium Access Control (MAC). TS DTS/SmartBAN(14)006001r5, Dec.

40. Salazar C, Kaiser A, Cathelin A and Rabaey J. 2015. A −97 dBm-sensitivity interferer-resilient 2.4 GHz wake-up receiver using dual-IF multi-N-Path architecture in 65 nm CMOS. *Proc. ISSCC*, pp. 396–398.

41. Tomabechi S, Komuro A, Konno T, Nakase H and Tsubouchi K. 2001. Design and implementation of spread spectrum wireless switch with low power consumption. *IEICE Trans. Fundamentals*, E84-A(4): 971–973.

42. Pletcher N, Gambini S and Rabaey J. 2007. A 65 μW, 1.9 GHz RF to digital baseband wakeup receiver for wireless sensor nodes. *Proc. CICC*, pp. 539–542. doi: 10.1109/CICC .2007.4405789.

43. Pletcher N, Gambini S and Rabaey J. 2009. A 52 μW wake-up receiver with −72 dBm sensitivity using an uncertain-IF architecture. *IEEE Journal of Solid-State Circuits*, 44(1): 269–280. doi: 10.1109/JSSC.2008.2007438.

44. Bryant C and Sjoland H. 2014. A 2.45 GHz, 50 uW wake-up receiver front-end with −88 dBm sensitivity and 250 kbps data rate. *European Solid State Circuits Conference (ESSCIRC)*, pp. 235–238. doi: 10.1109/ESSCIRC.2014.6942065.

45. Xiongchuan H, Rampu S, Xiaoyan W, Dolmans G and de Groot H. 2010. A 2.4 GHz/ 915 MHz 51 μW wake-up receiver with offset and noise suppression. *Proc. ISSCC*, pp. 222–223. doi: 10.1109/ISSCC.2010.5433958.

46. Joonsung B and Hoi-Jun Y. 2012. A 45 μW injection-locked FSK wake-up receiver for crystal-less wireless body-area-network. *Proc. A-SSCC*, pp. 333–336. doi: 10.1109/IPEC .2012.6522693.

47. Pandey J, Shi J and Otis B. 2011. A 120 μW MICS/ISM-band FSK receiver with a 44 μW low-power mode based on injection-locking and 9× frequency multiplication. *Proc. ISSCC*, pp. 460–462. doi: 10.1109/ISSCC.2011.5746397.

48. Hyunwoo C, Joonsung B and Hoi-Jun Y. 2013. A 37.5 μW body channel communication wake-up receiver with injection-locking ring oscillator for wireless body area network. *IEEE Trans. on Circuits and Systems I: Regular Papers*, 60(5): 1200–1208. doi: 10.1109/TCSI.2013.2249173.

49. Choi J, Lee I-Y, Lee K, Yun S-O, Kim J, Ko J, Yoon G and Lee S-G. 2014. A 5.8-GHz DSRC transceiver with a 10-μA interference-aware wake-up receiver for the Chinese ETCS. *IEEE Trans. MTT*, vol. 62, no. 12, pp. 3146–3160. doi: 10.1109/TMTT.2014 .2362118.

50. Moazzeni S, Cowan GER and Sawan M. 2012. A 28 μW sub-sampling based wake-up receiver with −70 dBm sensitivity for 915 MHz ISM band applications. *Proc. ISCAS*, pp. 2797–2800. doi: 10.1109/ISCAS.2012.6271891.

51. Moazzeni S, Sawan M and Cowan GER. 2014. An ultra-low-power energy-efficient dual-mode wake-up receiver. *IEEE Trans. on Circuits and Systems I: Regular Papers*, vol. 62 (2): pp. 517–526. doi: 10.1109/TCSI.2014.2360336.

52. Kuang-Wei C, Xin L and Minkyu J. 2012. A 2.4/5.8 GHz 10 μW wake-up receiver with −65/−50 dBm sensitivity using direct active RF detection. *Proc. A-SSCC*, pp. 337–340. doi: 10.1109/IPEC.2012.6522694.

53. Wada T, Ikebe M and Sano E. 2013. 60-GHz, 9-μW wake-up receiver for short-range wireless communications. *Proc. ESSCIRC*, pp. 383–386. doi: 10.1109/ESSCIRC.2013 .6649153.
54. Durante MS and Mahlknecht S. 2009. An ultra low power wakeup receiver for wireless sensor nodes. *Proc. STA*, pp. 167–170. doi: 10.1109/SENSORCOMM.2009.34.
55. Hambeck C, Mahlknecht S and Herndl T. 2011. A 2.4 μW wake-up receiver for wireless sensor nodes with −71 dBm sensitivity. *Proc. ISCAS*, pp. 534–537. doi: 10.1109/ISCAS .2011.5937620.
56. Nilsson E and Svensson C. 2013. Ultra low power wake-up radio using envelope detector and transmission line voltage transformer. *IEEE Journal on Emerging and Selected Topics in Circuits and Systems*, 3(1): 5–12. doi: 10.1109/JETCAS.2013.2242777.
57. Roberts NE and Wentzloff DD. 2012. A 98 nW wake-up radio for wireless body area networks. *Proc. RFIC*, pp. 373–376. doi: 10.1109/RFIC.2012.6242302.
58. Seunghyun O, Roberts NE and Wentzloff DD. 2013. A 116 nW multi-band wake-up receiver with 31-bit correlator and interference rejection. *Proc. CICC*, pp. 1–4. doi: 10.1109 /CICC.2013.6658500.
59. Semtech, SX1272 datasheet: SX1272/73—860 MHz to 1020 MHz Low Power Long Range Transceiver.
60. Axsem. AX5043 datasheet: Advanced high performance ASK and FSK narrow-band transceiver for 70-1050 MHz range.
61. Gartner. 2014. Gartner says the Internet of things installed base will grow to 26 billion units by 2020. Press release, Jan. 2.
62. ABI Research. 2013. More than 30 billion devices will wirelessly connect to the Internet of everything in 2020. Press release, May.
63. Texas Instruments, CC2420 datasheet: 2.4 GHz IEEE 802.15.4/Zigbee ready RF Transceiver.

Section II

Algorithms and Data Processing

6 Framework for Biomedical Algorithm Designs

Su-Shin Ang and Miguel Hernandez-Silveira

CONTENTS

6.1 INTRODUCTION

Designing effective engineering solutions or algorithms to solve biomedical problems in the healthcare monitoring context is a challenging task, to say the least. The initial objectives and specifications are typically vague. In addition to this lack of clarity, the designer has to take into account many factors, including the algorithmic complexity that can be tolerated by the target platform as well as the type and amount of data that are available to calibrate and properly evaluate the algorithm. For example, the physiological condition of a subject might adversely affect the signals, which might have unpredictable consequences on the algorithm used to process them. Therefore, limitations and issues with the algorithm might not be adequately exposed at design time, which could lead to substantial costs and loss of credibility when they are discovered by the client during or after field deployment.

Accurate assumptions are crucial during the design stages of the algorithm. While assumptions stated in prose are necessary and easily communicated across engineering and nonengineering staff, it is often imprecise and provides substantial scope for misinterpretation. Consequently, it is of the opinion of the authors that there should be a framework to formulate the problem and its constraints mathematically. The advantages of such an approach are twofold—first, it affords a level of clarity and precision that cannot be achieved by prose; second, it allows the performance of the candidate algorithm set to be quantified and objectively compared.

From actual problems that we encountered in the field, biomedical problems can be broadly divided into three types—optimization of a predefined objective function (Type 1), static classification problems (Type 2), and predictive models (Type 3). The first type of problems consists of trade-offs between different algorithmic factors that are reasonably well defined. Consequently, the objective function can be accurately stated. While rather rare, these types of problems do occur in the biomedical field, such as the compression algorithms for physiological algorithms. The second type of algorithms is extremely typical—the function used for classification purposes is unknown, but the data and the expected discrete outcomes (discrete decisions are required) are known, and the objective is to determine a classification function that would accurately map the initial data set to those observations. Finally, the last type of problems involves the prediction of the value of random variable/s, based on a set of observed features. There are substantial overlaps between each problem type, and an algorithm can frequently be used to solve multiple types of problems. However, they differ in terms of how the initial problem is formulated and consequently the design decisions that are taken.

In this chapter, we first provide an overview in terms of the problem types and how each can be practically formulated (Section 6.2). Since data are frequently crucial in the design process, we talk about how they can be acquired, and how the data can be used during the training and evaluation process, followed by a series of case studies, where the biomedical problem in question is formulated and solved (Section 6.3). Downstream of the design and prototyping process, the algorithm has to be ported to the target platform. We present such a design flow in Section 6.4. Finally, we conclude in Section 6.5.

6.2 FORMULATING A PROBLEM MATHEMATICALLY

From actual problems encountered in the field, we believe that they can be broadly divided into three types. The first type is the generic optimization problem (Type 1), which is used across multiple application domains. In this case, the objective or cost function and constraints are well defined. We discuss this at length in Section 6.2.1.

The second type of problem is the static classification problem (Type 2) involving discrete decisions, where the classification function is unknown, but the data set involving the original measurements and the discrete and expected outcomes are available (Section 6.2.2). The last type is prediction problems (Type 3), which require the prediction of random variables on the basis of the statistical distribution of related features (Section 6.2.3).

6.2.1 GENERIC OPTIMIZATION PROBLEMS (TYPE 1)

The problem that one is trying to solve is usually ill-defined in the initial stages. Examples include the derivation of optimum parameters for a compression algorithm, or the derivation of a hyperplane for accurate classification in the feature space. To elucidate the problem, the following steps can be followed:

1. Identify effective metrics used in quantifying how effectively a problem has been solved. This could mean the compression ratio and the quality of the reconstructed signal.
2. Identify the parameters with the most substantial impact on these metrics. In the context of signal compression, these parameters could include the quantization factors for the discrete cosine transform (DCT) coefficients.
3. If possible, define the problem formally as an optimization problem.

There are two aspects to an optimization problem—the objective function and the set of constraints. The optimization problem is defined over the parameter space, where each data point in the parameter space has a certain cost or benefit, which is quantified by means of an objective function. The scope of this parameter space is in turn defined by constraints, which are characterized by a set of equalities or inequalities. More formally, the objective function can be specified according to Equation 6.1.

$$\text{Minimize } f : \mathbb{R} \rightarrow \mathbb{R} \tag{6.1}$$

In this case, the objective function f is defined over a feasible set of parameters, $C \subset R_n$. The domain of the objective function is defined by a tuple with n real numbers—$\{x_1, x_2, ..., x_n\}$, and the range consists of single real numbers, which is the cost of using the corresponding set of parameters. The set of equalities and inequalities involving the constraint functions, used to define the scope of the parameter space, are shown in Equations 6.2 through 6.4.

$$f(x) \le c_i \tag{6.2}$$

$$f(x) \ge c_j \tag{6.3}$$

$$f(x) = c_k \tag{6.4}$$

In certain cases, it might be useful to convert the inequalities to equalities using slack variables, s_i, to enable certain methods such as Integer Linear Programming [1] to be employed (Equation 6.5).

$$g(x) = c_i + s_i \tag{6.5}$$

A minimization problem can easily be converted to a maximization problem by negating the objective function. If the functional surface is convex with a single saddle point, the set of parameters corresponding to this point represents the best solution to this problem. In addition, for many problems, the functional surface has a large number of saddle points (or local minima), and it may not be feasible to carry out an exhaustive search to explore all candidate solutions in order to arrive at the globally optimum solution.

Ideally, the problem definition should be independent of the algorithm. However, more often than not, the parameters concerned are dependent on the algorithm used. In this case, it is necessary to redefine the problem for each algorithm in the candidate set. The metrics used in determining the outcome should remain uniform, so that the performance of different algorithms can be properly compared.

6.2.2 STATIC CLASSIFICATION PROBLEMS (TYPE 2)

A specific type of problem that typically occurs in the biomedical field is classification, where a discrete decision needs to be made—given a measurement, is it possible to *automatically* predict an outcome with a high degree of accuracy? For example, one might like to build a mechanism to identify abnormal patterns within electrocardiograms (ECGs) in order to determine if a patient is suffering from an arrhythmia. Frequently, the measurements and the expected outcomes (expert opinion by a clinician) are available, but the classification function is unknown and needs to be determined.

More formally, the initial data set, D, consisting of the measurements, is provided. This data set contains a set d tuples, $\{t_0, t_2, ..., t_{d-1}\}$. In order to reduce the dimensionality of the problem, feature extraction is carried out in order to extract the most pertinent aspects of the data. Therefore, each tuple t_i is associated with feature vector $V_i = \{f_{i1}, f_{i2}, ..., f_{iN}\}$, containing N features. Specifically, $V_i \subset V$, where V is known as a feature matrix. In addition, a corresponding class label C_i should also be available for each t_i and V_i, and $C_i \subset C$ (C is known as the observation vector). The objective is therefore to derive a mapping function, f, from V to a predicted set of outcomes, P, where the difference between P and C is minimized.

A class of solutions, known as supervised machine learning algorithms, has been designed with this type of problems. The steps involved in problem solving are as follows [2]:

1. *Feature extraction*: Determining and extracting the most relevant aspects from the raw data, in order to facilitate more effective classification.
2. *Feature selection*: Select a subset of the most relevant features from the original set (columns of feature matrix V). This step can be carried out either manually or automatically.
3. *Feature space mapping*: To improve classification accuracy, a transformation is carried out on the original to a new feature space. This might involve increasing or decreasing the dimensionality of the feature space.
4. *Classifiers*: There are many different types of classification techniques, with differing types of classification functions (hyperplanes within the feature space).
5. *Validation/testing*: This phase involves the methodologies in the training and testing of the classifiers with two objectives—optimizing the parameters of the classifier and obtaining representative metrics for the performance of the classifier.

This supervised machine learning framework takes into account human input and experience at the initial stage, specifically at the feature extraction stage. On the basis of a combination of experience and trial and error, the initial set of features that are likely to enable effective classification are identified. For example, patients suffering from a typical arrhythmia, atrial fibrillation (AF) [3], is known to exhibit substantial variability in the instantaneous heart rate (HR). In consequence, the root mean square of the successive differences of peak-to-peak intervals (RMSSD of the times between each heartbeat) is often considered as a candidate feature. Once the features have been identified, the feature space may be described, as seen in Figure 6.1 [4]. This two-dimensional feature space is populated by two classes (marked differently in the feature space), which are concentrated in distinct clusters, indicating that there is good separability between the classes in this space. The lines separating these clusters are defined by classification functions, or hyperplanes, which partition the feature space into independent components, so that any new incoming data may be automatically classified on the basis of its location within this space.

Redundant and irrelevant features may confound the classification problem. Consequently, it is necessary to cull them from the final feature set. The most straightforward manner is to do so manually, by visually assessing the separability of the classes in the feature space. There are two limitations to this approach—first, visual assessment is subjective and may therefore be inconsistent between different designers. Second, this approach is limited to at most three dimensions. Potentially, automatic feature selection techniques may be used to overcome these limitations. Two methods that are used for this purpose are the Minimum Redundancy Maximum Relevance (mRMR) [5] and the RELIEFF [6] algorithms. The mRMR algorithm removes redundant features by locating a subset of features that minimizes the metric, mutual information (normalized form of correlation) between different features, and maximizing the same metric between selected features and the observation vector. The RELIEFF algorithm takes a different approach. In particular, it considers the distance of each feature vector from its two nearest neighbors from the same and the opposite class (assuming a two-class problem) [6], penalizing the feature if it is near the companion in the same class and vice versa for the nearest vector in the opposite class.

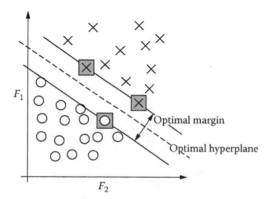

FIGURE 6.1 An example of a feature space and the corresponding hyperplane, derived from the support vector machine. F_1 and F_2 are two different features. (From C. Cortes and V. Vapnik, *Machine Learning*, (3):273–297, 1995.)

Similar to feature selection, the objective of the feature mapping stage is to maximize separability between different classes in the feature space. This can be achieved by increasing the dimensionality of the feature space by means of techniques such as kernel substitution [7] or dimensionality reduction techniques such as Sammon mapping [8]. In the case of kernel substitution [7], an extra dimension in the feature space is created, and this has been shown to improve classification performance in some applications. An example is $K(x,x') = x^T x$. Another popular kernel is the radial basis function, where the transformed feature space contain Euclidean distances between feature vectors, rather than the absolute positions of the vectors themselves.

After the best features and transformation function have been selected for the problem, different classification methods may be applied to the problem. These methods result in different hyperplanes, and the best classifier is dependent on the data set as well as the problem. The best known classifier is the artificial neural network (ANN), as shown in Figure 6.2. This consists of the basic node or neuron, which is composed of a linear function of adjustable weights and the inputs from other nodes. The output of this function is then fed into a smooth differentiable sigmoidal function. These nodes are organized into the input layer, hidden layers, and the output layer. The number of nodes in the input and output layers is bounded by the number of features and the classification problem at hand, respectively, whereas the number of hidden layers and the number of nodes in each hidden layer are design parameters. In general, a large number of layers and nodes in the hidden portion afford higher degrees of freedom and, subsequently, a more versatile and accurate hyperplane. On the other hand, it could lead to the problem of overfitting, or a nongeneric hyperplane. Once the topology of the ANN has been fixed, the weights are determined by a technique known as error back propagation [9]. Essentially, training data vectors are fed to the ANN during the forward phase of the algorithm. Subsequently, during the backward phase, the error between the label and the current ANN output is "back-propagated" and the weights of each node are adjusted using the gradient

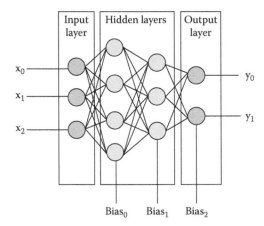

FIGURE 6.2 A back-propagated ANN, with input x and output y. All of these nodes contain adjustable weights, in order to minimize the errors between the y and the expected outcomes.

descent method. With enough representative data, the weights will eventually converge to values where the classification error is minimized.

Two other popular machine learning techniques are decision trees and support vector machines (SVMs). An example of a decision tree is furnished in Figure 6.3, which is designed for arrhythmia detection, on the basis of features extracted from a segment of ECG data—RMSSD in beat-to-beat intervals and the average corrected (using an ectopic filter) beat-to-beat interval. Each node of the tree consists of an inequality, with the feature and a threshold. These parameters are obtained by means of the C4.5 algorithm [10], where the feature matrix and observation vector are recursively divided at each node, on the basis of the best feature and the optimal threshold found for that feature and that data subset. The metric used in selecting the feature for each node is the information gain [10]. The SVM is another popular machine learning algorithm that has grown in popularity over the recent years. A hyperplane that is derived from this approach can be seen in Figure 6.1. Essentially, this technique derives the optimal hyperplane by maximizing the margin between the data points (support vectors) closest to the hyperplane. Kernel substitution is frequently embedded into the objective function, used in the derivation of the hyperplane, in order to improve classification performance.

As one might expect, the parameters of the classifier are dependent on the data used in the training process. In addition, the initial classifier state and the manner in which data are fed during the training process has an impact as well. These issues will be discussed in Section 6.3. Once the classification parameters are determined, they cannot be changed after the training stage. When the expected outcomes or the observation vector are unavailable, unsupervised machine learning algorithms may be used to detect clusters within the underlying data structure in order to determine the labels. However, we consider this class of problems as outside the scope of our discussion.

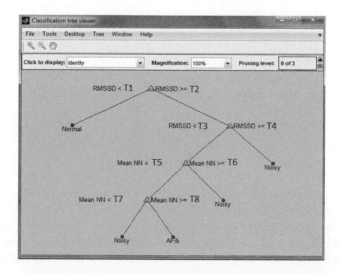

FIGURE 6.3 A binary decision tree for arrhythmia classification, using the features RMSSD and mean NN (corrected beat-to-beat intervals). T1 to T8 represent thresholds derived from the C4.5 algorithm. (From J. R. Quinlan, *C4.5: Programs for Machine Learning.* Morgan Kaufmann Publishers, USA, 1993.)

6.2.3 Prediction Problems (Type 3)

In Section 6.2.2, the required outcomes are discrete and the parameters of the classification model are fixed after the training stage and cannot be altered during runtime. Therefore, it is very important that the true distribution of the feature space is understood and well represented by the data. Frequently, this is not the case. For instance, if one is trying to determine the reliability of HR values produced by an algorithm, the type of artifacts produced by bedridden patients in a general ward would differ from those produced by frequently ambulating patients in the accidents and emergency department, prompting the need for a classifier that is capable of adjusting to the incoming data distribution. Potentially, if the training data set originates from a single source, it could result in a classifier that is customized to the physical location of the patients as well as their physiological conditions.

Assuming that one is trying to predict the value of a nondeterministic random variable, C, which is dependent on a set of observed features, $\{F_j | j = 1, ..., M\}$. By the Bayes theorem, the conditional probability distribution of C is a function of the prior class distribution, $P(C)$, and the likelihood function, $P(F_1, ..., F_{M-1}|C)$, as shown in Equation 6.6.

$$P(C|F_1,...,F_M) = \frac{P(C)P(F_1,...,F_M|C)}{P(F_1,...,F_M)} \tag{6.6}$$

The posterior probability can be factorized and reexpressed using the chain rule, as shown in Equation 6.7. By comparing Equation 6.6 with Equation 6.7, it follows that the likelihood function can be expressed as Equation 6.8. If it can be further assumed that the observed features are independent, Equation 6.8 reduces to Equation 6.9.

$$P(C|F_1,...,F_M) = P(C)P(F_1|C)P(F_M|C, F_1,...,F_{M-1}) \tag{6.7}$$

$$P(F_1,...,F_M|C) \propto P(F_1|C_i)P(F_2|C, F_1)P(F_M|C, F_1,...,F_{M-1}) \tag{6.8}$$

$$= \prod_{j=1}^{M} P(F_j|C) \tag{6.9}$$

We carry on by making the assumptions that the prior and likelihood distributions are Gaussian. Further, N independent observations of $\mathbf{F}$ were made ($\mathbf{F} = \{F_1, F_2, ..., F_M\}$); hence, the resulting likelihood function can be formulated, as shown in Equation 6.11. It is assumed that the mean and standard deviation of F are given by μ and σ, respectively. Similarly, the prior distribution is described in Equation 6.12.

$$P(\mathbf{F}|C) \propto \prod_{n=1}^{N} P(\mathbf{F}_n|C) \tag{6.10}$$

$$\propto \frac{1}{(2\pi\sigma^2)^{\frac{N}{2}}} \exp\left\{-\frac{1}{2\sigma^2}\sum_{n=1}^{N}(F_n-\mu)^2\right\} \tag{6.11}$$

$$P(C) : N\left(\mu_0, \sigma_0^2\right) \tag{6.12}$$

Given that the posterior distribution $P(C|F)$ is a function of both the prior and likelihood function, it is also Gaussian (represented by $N\left(\mu_N, \sigma_N^2\right)$). With each incoming datum, both the prior probability and the likelihood functions can be sequentially updated—the likelihood function, $P(F|C)$, is updated with each new data point. This is known as the inference step. Subsequently, the prediction step is made by taking the product between the prior probability and the likelihood probability (decision step), to obtain the posterior probability. The posterior probability can be reexpressed as shown in Equation 6.14. The term in the square bracket is the posterior probability from the previous step $(N-1)$, and this is resubstituted as the prior probability in the subsequent time step. Clearly, this is an iterative process, with the inference and decision step taking placing cyclically.

$$P(F|C) \propto P(C)\prod_{n=1}^{N}P(F_n|C) \tag{6.13}$$

$$= \left[P(C)\prod_{n=1}^{N-1}P(F_n|C)\right]P(F_N|C) \tag{6.14}$$

Further, it can be shown that the mean and variance of the resulting distribution are given by Equations 6.15 and 6.16, respectively [7]. The parameter, μ_{ML}, refers to the mean value derived from maximizing the likelihood function. From Equation 6.15, it can be seen that for a small number of observations, N, the prior mean has a bigger influence over the posterior mean, compared to the maximum likelihood mean value, and vice versa. In addition, from Equation 6.16, it is clear that the initial variance is high but falls in magnitude as N grows. This observation implies that the amount of uncertainty in the prediction of C decreases with more incoming data.

$$\mu_N = \frac{\sigma^2}{N\sigma_0^2+\sigma^2}\mu_0 + \frac{N\sigma_0^2}{N\sigma_0^2+\sigma^2}\mu_{ML} \tag{6.15}$$

$$\frac{1}{\sigma_N^2} = \frac{1}{\sigma_0^2} + \frac{N}{\sigma^2} \tag{6.16}$$

Thus far, it has been assumed that both the features and, correspondingly, the predicted variables are independently and identically distributed. In many applications, this assumption is unrealistic, and more accurate predictions may be facilitated by capturing the correlation between the predicted variables themselves. For instance, one might be trying to predict if an abnormal rhythm is detected from the ECG of a patient. The most recent prediction is likely to have a bearing on the current prediction, as rhythms tend to be persistent. Potentially, we can capture this temporal correlation using a state space representation of the system. In this paradigm, there are two parameters—observations and latent variables. They correspond to the observed features ($\mathbf{F_i}$) and the parameter that we are trying to predict (C_i, which is assumed to be a discrete random variable in this case), respectively. This particular state space representation is known as a hidden Markov model (HMM), and it is illustrated in Figure 6.4. Note the edges between consecutive values of C_i, which represent the dependence between them.

In the case of the HMM, its joint distribution is shown in Equation 6.17 [7]. The conditional parameters for this distribution are captured by θ, which, in turn, contains $\{\phi, A, \pi\}$, which are explained as follows. ϕ represents the parameters governing $\mathbf{F}$, conditioned upon the state, C. For instance, if $\mathbf{F}$ is a continuous random variable that is normally distributed, ϕ would represent the mean and the standard deviation of the distribution. Each element of the transition matrix, A_{ij}, situated in row i and column j, represents the probability that the system will transit to state j, given that it is originally in state i. Assuming that C can take any one of K states, the vector π will be populated by the probabilities of the system being in a particular state (for instance, $\pi = \{P(\text{Abnormal rhythm detected}), P(\text{Normal Sinus Rhythm})\}$).

$$P(\mathbf{F}, C | \theta) = P(\pi) \left[\prod_{n=2}^{N} P(C_n | C_{n-1,A}) \right] \prod_{m=1}^{N} P(\mathbf{F}_m | C_m, \phi) \qquad (6.17)$$

$\theta = \{\pi, A, \phi\}$
$\pi = $ Probability of the system taking a certain state
$A = $ Transition matrix, $A_{i,j} = P(C_t = j | C_{t-1} = i)$
$\phi = $ Statistical parameters for the emission distribution

A log-likelihood function can be formed using the "old" posterior marginal (based on data governed by previous statistical parameters θ^{old}) and the logarithm of the present marginal distribution, as shown in Equation 6.18 [7]. Equation 6.18 can then be combined with Equation 6.17 to obtain Equation 6.19. In this new expression, two new parameters—the marginal posterior probability [$\gamma(C_n)$] and the joint posterior probability of successive latent variables [$\xi(C_{n-1}, C_n)$] are introduced. By maximizing the likelihood function with respect to θ, it can be shown that the optimal solutions for π and A are functions of γ and ξ [7].

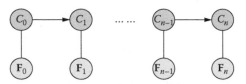

FIGURE 6.4 State space diagram for the HMM.

$$Q(\theta, \theta^{\text{old}}) = \sum_Z P(C|\mathbf{F}, \theta^{\text{old}}) \ln P(C, \mathbf{F}|\theta) \tag{6.18}$$

$$Q(\theta, \theta^{\text{old}}) = \sum_{k=1}^{K} \gamma(C_{1k}) \ln \pi_k + \sum_{n=2}^{N} \sum_{j=1}^{K} \sum_{k=1}^{K} \xi(C_{n-1,j}, C_{n,k})$$
$$+ \sum_{n=1}^{N} \sum_{k=1}^{K} \gamma(C_{nk}) \ln P(\mathbf{F}_n|\phi_k) \tag{6.19}$$

$$\gamma(C_n) = P(C_n|\mathbf{F}, \theta^{\text{old}}) \tag{6.20}$$

$$\xi(C_{n-1}, C_n) = P(C_n, C_{n-1}|\mathbf{F}, \theta^{\text{old}}) \tag{6.21}$$

The marginal posterior probability, γ, and the joint posterior probability of successive latent variables, ξ, are in turn derived as a function of the forward and backward conditional probabilities—α and β, respectively, based on the Bayes theorem, which is seen in Equation 6.22. At time step n, the forward probability is the joint probability of the current prediction and all past observations, while the backward probability is the probability of all future observations, conditioned upon the current prediction. These parameters are respectively described in Equations 6.24 and 6.25. Note that they are recursively expressed and are therefore updated at every time step. The additional parameters, $P(F_i|C_j)$ and $P(C_i|C_j)$, are known quantities in each time step and can be derived from θ.

$$\gamma(C_n) = P(C_n|\mathbf{F})$$
$$= \frac{P(\mathbf{F}|C_n)P(C_n)}{P(\mathbf{F})} \tag{6.22}$$
$$= \frac{\alpha(C_n)\beta(C_n)}{P(\mathbf{F})}$$

$$\xi(C_{n-1}, C_n) = P(C_{n-1}, C_n|\mathbf{F})$$
$$= \frac{\alpha(C_{n-1})P(F_n|C_n)P(C_n|C_{n-1})\beta(C_n)}{P(\mathbf{F})} \tag{6.23}$$

$$\alpha(C_n) = P(\mathbf{F}_1, \ldots, \mathbf{F}_n, C_n)$$
$$= P(\mathbf{F}_n|C_n) \sum_{C_{n-1}} \alpha(C_{n-1})P(C_n|C_{n-1}) \tag{6.24}$$

$$\beta(C_n) = P(\mathbf{F}_{n+1}, \ldots, \mathbf{F}_N|C_n)$$
$$= \sum_{C_{n+1}} \beta(C_{n+1})P(F_{n+1}|C_{n+1})P(C_{n+1}|C_n) \tag{6.25}$$

Clearly, future data are required for HMMs to function, as seen in Equation 6.25, which may be unavailable for certain applications. The alternative is to make use of predictors such as Kalman filters [11], which would require only present and historical data. This predictor and its application will be described in Section 6.3.6. Note that the above description of the HMM is by no means comprehensive and readers are referred to Ref. [7] for more details.

6.3 CHARACTERISTICS OF BIOMEDICAL PROBLEMS

6.3.1 DATA TYPES

As mentioned in Section 6.2, data are crucial for both the design and evaluation of the algorithms. In particular, for classification and prediction problems, the results are closely linked to the initial data set fed to the training algorithms. Since there is no way to constrain the problem directly, the indirect option is to populate the training and testing with carefully specified data. Some suggested guidelines are as follows:

1. *Physiological artifacts*: The data set should consist of signals reflecting the extreme and typical physiological state of the patients. For example, in the context of arrhythmia detection, signals resulting from different types of arrhythmias (apart from the one that the classifier is trying to detect) should be taken into account. This would include tachycardia (abnormally high HR), bradycardia (abnormally low HR), and various ventricular and atrial arrhythmias.
2. *System artifacts*: Noise is a common occurrence in biomedical signals. Sources originating from the system would include motion hardware noise, mains noise, and artifacts introduced by front-end sensors or other system components such as the radio. These confounding artifacts should be taken into account in the design of the data set.
3. *Mechanical artifacts*: Mechanical movements on the part of the subjects could introduce motion artifacts in the acquired physiological signals and confound the measurements. Clearly, this is dependent on the type of motion, which is difficult to predict comprehensively during design time. By far, this type of artifacts is the most difficult to characterize.
4. *Prior data distribution*: The distribution of different types of signals should be clearly specified, reflecting that of the target population.

The amount of data available depends on the application being considered. For applications such as ECG signal compression or arrhythmia detection, there are several online repositories [12], which are freely available to algorithm designers. However, these signals are often acquired using front-end sensors, different from those used to collect data for the target application. Consequently, artifacts particular to the target system cannot be accurately captured.

Synthetic data generators, generating particular types of physiological signals, are available. On the other hand, most such tools do not accurately capture the above-mentioned artifacts. To resolve this problem, a new tool or framework could be used for synthetic signal generation. As an input, the tool could take in a set of "seeds" or signals collected from available subjects in a controlled setting (e.g., in a laboratory). Motion artifacts could then be independently collected by mounting the target sensor on a rig and subjecting it to representative movements that the subject is likely to engage in, such as walking, eating, and talking. In addition, underlying artifacts could be extracted from the seeds using predictors such as Kalman filters [11] to obtain the residual noise. These independently collected artifacts can then be superimposed on top of the original seeds to generate further signals. Directly adding physiological artifacts to the signal set could be achieved by means of patient simulators. However, the type of rhythms and the available set of patterns are often limited. Consequently, morphological arrhythmic transformation operators can be applied to the seeds in order to generate signals with abnormal rhythms. For example, the operator should be able to transform the signal from one shown in Figure 6.5a [13] to that in Figure 6.5b [14]. These signals should be convincing enough to "fool" expert clinicians.

Apart from the aggregate classification metrics used in assessing classification accuracy (positive predictivity [+P], sensitivity [Se], and specificity [Sp] [2]), it is pertinent to accurately specify information about the type of data used in the training and evaluation of the data. In addition, tests for statistical significance should be carried out in order to determine differences between the reference and the algorithmic results. If the distributions of both the reference and the algorithmic results data sets are Gaussian, the Student t test [15] should be used to test the statistical significance. Otherwise, nonparametric methods [15] should be used instead. Finally, the differences between the data sets should also be quantified using correlation analysis, as well as the 95% confidence and prediction intervals.

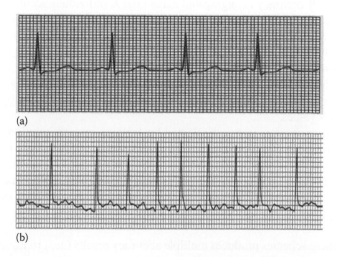

(a)

(b)

FIGURE 6.5 Examples of ECGs from (a) a healthy patient (From S. Navas, *Nursing Standard*, (17):45–54, 2003) and (b) a patient suffering from AF (From S. Goodacre and R. Irons, *British Medical Journal*, (324):594–597, 2002).

6.3.2 Training Frameworks

The classifier parameters are dependent on the data used during the training stage, as well as the manner in which the data are fed and used for evaluation. If a high degree of correlation exists between consecutive data vectors, the classifier parameters might get stuck in a local minimum [9]. This suggests that the arrangement of data vectors ought to be randomized before they are being fed for training and evaluation purposes. This is particularly important if the primary objective is to evaluate the relative performance of different classifiers. The collected data set is partitioned into subsets of two types—training and testing. The metric used to determine the accuracy of a classifier is shown in Equation 6.26, for a data set populated with h data vectors. In most methods, this is a subset of the original data set of size N.

$$acc_h = \frac{1}{h} \sum_{P_i, C_i} \delta(P_i, C_i) \tag{6.26}$$

There are many schemes used for training and testing classifiers. Some of these schemes are as follows [16]:

1. *Holdout*: This involves partitioning the data set into mutually exclusive training and testing (or holdout) sets and subsequently obtaining the classification accuracy using Equation 6.26. This is done iteratively in order to obtain the mean accuracy.
2. *K-fold cross validation*: This involves partitioning the data set into K folds or equally sized subsets, using $K - 1$ sets for training, and the remaining set as a test set for accuracy evaluation. This is done iteratively K times, and the overall accuracy is aggregated across the K individual accuracy values that are obtained. A stratified version of this scheme involves adjusting the distribution of each fold such that it has the same distribution as the original data set.
3. *Leave one out*: This involves leaving one data vector for testing purposes and using the remaining $(N - 1)$ vectors for training purposes, and doing this iteratively N times. In this case, the overall accuracy is aggregated across N values.
4. *Bootstrap*: one data vector is selected for training in each iteration, and the remaining data vectors are used for testing the classifier. This is done b times in order to obtain a set of b accuracies, $\{\varepsilon_i | i = 1, ..., b\}$. In addition, the resulting classifier is applied to the training set to obtain an accuracy value, acc_s, and the overall bootstrap accuracy is obtained as a function of both ε_i and acc_s.

Each of these schemes produces multiple accuracy results (acc_h in Equation 6.26), which can then be aggregated to obtain the mean error and variance. The ideal scheme would result in a mean error of zero (or low bias) and low variance in errors. According to different reviews, these parameters were dependent on the number of data vectors and the number of features [16]. There was agreement that stratified 10-fold cross validation was found to produce favorable results [16].

The work mentioned so far makes use of all of the data in the training set. However, for ANNs, it was found that a more representative classifier (lower risk of overfitting) can be obtained by considering the distribution of the neural network weights, **w** [17]. Therefore, given the data set D, the set of weights chosen at each time instance can be modeled as a random variable, **w**. Therefore, the objective is to sample from the posterior probability, $p(\mathbf{w}|D)$ using the Metropolis algorithm [18], retaining or discarding each **w** heuristically in each iteration, in order to converge near/at an optimum solution. On the other hand, the Metropolis algorithm is often found to converge much more slowly; thus, the hybrid Monte Carlo algorithm is used instead [17]. These techniques could be used when direct usage of the data for training causes problems of overfitting owing to the dearth of data. However, training times are often much longer in such cases, compared to the conventional training methods described above [17].

6.3.3 Case Study 1: Compression of Physiological Signals (Type 1)

In the context of designing an ambulatory health monitoring device for patients, it is advantageous for the acquired physiological signals to be made available to the attending nurse or doctor, so that abnormalities detected within the signals can trigger early clinical actions such as comprehensive tests or measurements (e.g., a 12-lead ECG). High bandwidth is necessary for the streaming of multichannel physiological signals, and it is necessary to compress the signal in order to meet specifications for energy consumption for the product. Clearly, the compression algorithm itself has to be sufficiently energy efficient so as not to negate the benefits from compression.

In this case study, we consider the use of Type II DCT transform for the compression of ECG signals, which are shown in Equations 6.27 and 6.28, respectively, and where block size is N. This transform is used in JPEG compression [19]. In addition, it demonstrated competitive compression ratios and distortion errors for physiological signals as well [20]. Specifically, the original ECG signal has a high level of redundancy and spatial correlation. This transform operates on a block-by-block basis and compresses the energy of the signal into the lower-frequency bins of the transform. Alternatively, the fast Fourier transform could be used, but it produces a complex output spectrum for a real signal and is therefore more compute intensive. On the other hand, the DCT produces only real coefficients for a real input signal, as shown in Figure 6.6. While the original signal is observed to vary relatively slowly, signal fluctuations are observed to occur at just the low-frequency bins of the spectrum and extremely low amplitude components (dominated by zeros) toward the higher end of the spectrum.

$$F(k) = \sum_{j=0}^{N-1} f(j)\cos\frac{(2j+1)k\pi}{2N} \qquad (6.27)$$

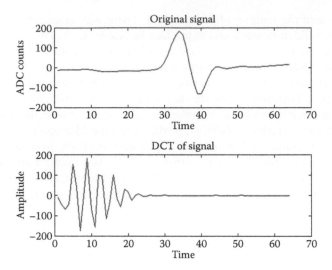

FIGURE 6.6 The top chart shows a signal segment containing an ECG QRS complex, while the chart at the bottom shows the compacted spectrum of the DCT.

$$f(j) = \sum_{k=0}^{N-1} c(k)F(k)\cos\frac{(2j+1)k\pi}{2N}$$

(6.28)

$$j,k = 0,\ldots,N-1$$

$$c(k) = \begin{cases} \dfrac{1}{\sqrt{(N)}}, & k = 0 \\[2ex] \sqrt{\dfrac{2}{N}}, & \text{otherwise} \end{cases}$$

The flowcharts in Figure 6.7 show the encoder and decoder used for the compression and reconstruction of the ECGs. This algorithm is a lossy compression algorithm—the compression process introduces distortions within the reconstructed signal. Specifically, the lossy aspect of the algorithm is introduced by the quantization block in Figure 6.7a. Because of the relative lack of sensitivity of the least significant bits of the DCT coefficients to the final quality of the reconstructed signal, they may be truncated or quantized to expose further redundancy within the signal. Downstream of the quantization module, a lossless compression algorithm—arithmetic compression [21]—is used to remove the redundancy within the sequence of quantized DCT coefficients. Using this lossless compression scheme is advantageous because the parameters of the algorithm are dynamically adjusted to the

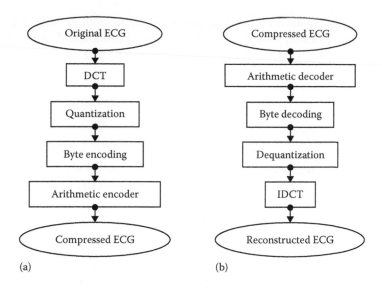

FIGURE 6.7   (a) DCT-based encoder. (b) DCT-based decoder.

immediate distribution of the signal, on the basis of incoming samples. As can be seen in Figure 6.7b, the decoder is symmetric to the encoder.

In order to determine the best trade-off between the compression ratio and the amount of distortion in the reconstructed waveform, the parameters within the quantization module may be adjusted. Each DCT block is modified according to Equation 6.29. The operator "//" refers to a division operation followed by rounding.

$$F(n) = \begin{cases} 0, & F(n) < t(n) \\ F(n)//q(n), & \text{otherwise} \end{cases} \qquad (6.29)$$

The set of quantization factors are defined by $q(n)$, $n = 0, ..., N - 1$ for a block size of N. Since frequency components of low amplitudes have little impact on the level of distortion, a separate set of thresholds, $t(n)$, $n = 0, ..., N - 1$, is used to remove them. The objective is to determine optimum values for these parameters where both compression ratio and reconstruction signal quality are maximized. This problem can be formulated as a Lagrangian [20], as shown in Equation 6.30, where J_n is to be minimized. Entropy is a measure of how effectively data can be compressed, and it is expressed as bits per symbol. To compute entropy, the distribution of the data (in this case, DCT coefficients) has to be available.

$$J_n = H_n(q(n), t(n)) + \lambda D_n(q(n), t(n)) \qquad (6.30)$$

H_n = Entropy of DCT coefficient n
D_n = Mean squared error of coefficient n
λ = Lagrange multiplier

This is easily obtained by constructing a histogram from existing data. From the histogram, the entropy can be determined using Equation 6.31, where p_i is the approximate probability of the occurrence of value i, $0 \leq i \leq M - 1$.

$$H = -\sum_{i=0}^{M-1} p_i \log_2 p_i \qquad (6.31)$$

The Lagrangian problem is graphically illustrated in Figure 6.3. Each point in the trade-off curve represents an optimal trade-off, on the basis of the specified level of distortion and entropy. For a specific value of λ, a corresponding trade-off occurs, which is tangential to the trade-off point, as illustrated in Figure 6.8.

Subsequently, the bounds of the algorithmic parameters have to be defined. Intuitively, more compaction can be achieved with a higher DCT block size. However, the complexity of the DCT algorithm varies quadratically with block size, $O(N^2)$, and compaction gains from the transform saturate at a block size of 64 [20]. Therefore, the DCT block size is set to 64. Similar observations were found for the quantization factors and the thresholds. Therefore, the upper bounds for them are set at 64. The lower bound for the threshold is set at $q(n)/2$ since the least significant bits of the corresponding coefficient will be removed during quantization, and the resolution is set to 0.5. Consequently, the bounds on the algorithmic parameters are shown in Equations 6.32 through 6.34.

$$q(n) = 1,\ldots,N \qquad (6.32)$$

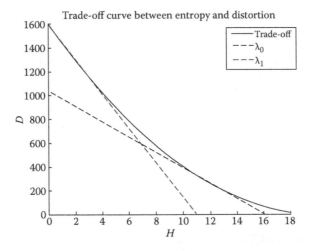

FIGURE 6.8   Illustration of the Lagrangian trade-off curve.

$$t(n) = \frac{q(n)}{2}, \frac{q(n)+1}{2}, \ldots, N - \frac{1}{2}, N \qquad (6.33)$$

$$N = 64 \qquad (6.34)$$

Obtaining the trade-off curve exhaustively will take a long time because of the large number of combinations of quantization factors, q, and thresholds, t, for each DCT coefficient. In addition, each data point in the trade-off curve has to be computed over a sufficiently large data set to obtain a representative cost of that particular trade-off. For these reasons, heuristics are used to reduce the time necessary to converge on an optimal solution, for different values of λ, from 0.1 to 1, in steps of 0.1. The simplex algorithm [22] is used to obtain the best trade-off. It is used for the following reasons:

1. The optimization algorithm avoids the local minimum by periodically expanding the solution space.
2. When moving toward the optimum solution, it takes big steps and is therefore likely to converge on the optimal solution in a smaller amount of time.

For each value of λ, the solution space, J, can be described by means of a polytope, where each data point is computed for unique values of q and t. The boundaries of the polytope are defined by its vertices. As the algorithm progresses, the polytope gradually "migrates" to a region where the global minimum occurs, and shrinks in size, as it converges around the optimum solution. At each step of the optimization process, the vertices are ranked according to its cost, and the vertex with the lowest cost is returned at the end of the optimization process. There are a few phases in the optimization algorithm, as follows:

1. *Reflection*: The vertex with the highest cost is reflected about the centroid of the polytope and replaced if the new vertex has a lower cost than the old vertex.
2. *Expansion*: With reference to the centroid, the polytope is expanded in the direction of the vertex with the lowest cost, on the basis of the heuristic that the "good" solutions are likely to occur in proximity.
3. *Contraction*: The polytope is shrunk during more mature stages of the optimization process.
4. *Reduction*: The polytope is shrunk around the vertex with the lowest cost, as the optimization process converges rapidly around the optimum solution.

From this optimization process, a set of N quantization factors and thresholds is obtained. These values are stored in the encoder and applied to each coefficient before the application of the next compression stage. The same quantization factors, $q(n)$, are stored in the decoder and used in the recovery of the coefficients. Details concerning the byte encoding/decoding and the arithmetic encoder/decoder modules are outside the scope of this chapter. Readers are referred to Ref. [20] for further information.

6.3.4 Case Study 2: Estimating Calorie Energy Expenditure (Type 1)

Studies have shown that a sedentary lifestyle and other unhealthy lifestyle choices are linked to the occurrence of type 2 diabetes [23]. According to the World Health Organization, 11% of the population is suffering from this disease, and this problem is imposing a substantial burden on national healthcare systems worldwide [24]. Through changes in lifestyle, including regular exercise, it was suggested that these risk factors could be reversed [23]. With the development of low-power physiological and biomechanical sensors, lightweight wearable devices can be used for the estimation of calorie energy expenditure of human subjects. These quantitative measures would allow the consulting doctor to determine if the patient is exercising at the prescribed level, or whether improvements have been made from the previous consultation.

A highly accurate technique of measuring calorie energy expenditure is achieved through the use of indirect calorimetry (an example of an indirect calorimeter is shown in Figure 6.9). This method is premised on the fact that energy is required for muscular contraction in order for physical activity to occur. The release of energy occurs through the metabolism of fat, carbohydrates, and proteins, which requires oxygen and produces carbon dioxide as one of the by-products. Therefore, by allowing the subject to breathe in air of known concentration (20.9% oxygen, 0.03% carbon dioxide, 79.1% nitrogen), and measuring the concentration of gases in exhaled air by means of a face mask, the oxygen uptake and the expelled carbon dioxide can be determined. The calorie energy expenditure can then be worked out as a function of oxygen uptake and carbon dioxide production (Weir formula [25]). Other accurate techniques for measuring calorie energy expenditure include direct calorimetry [26] and the doubly labeled water technique [27].

The merits of indirect calorimetry include its accuracy as well as its capability of measuring calorie expenditure on a breath-by-breath basis. However, the need for a face mask and its obtrusiveness precludes it as a device for use in daily living. These disadvantages motivated the development of models for the estimation of calorie expenditure using less obtrusive physiological and biomechanical sensors. Indeed, the HR is frequently used as a feature for energy expenditure prediction as it is a good indicator for cardiovascular stress [28]. Despite the linear relationship between HR and oxygen consumption at certain levels of exercise (particularly low and moderate), this parameter should not be used by itself as an indication for energy expenditure. The reason behind this is that HR is also influenced by factors other than activity, that is, emotional stress, anxiety, level of fitness, type of muscular contraction, active muscle group, environment, and hydration [29].

Another means of estimating energy expenditure involves the use of motion sensors such as accelerometers. This is based on the fact that energy expenditure increases proportionally with the muscular activity responsible for acceleration and movement of the body and its extremities during physical exercise and locomotion. Recent advances of microengineering technologies have enabled the development of small portable and wearable devices intended for measuring physical activity. These lightweight and unobtrusive systems are equipped with accelerometers and data-logging capabilities. Therefore, they can be used in routine clinical practice and

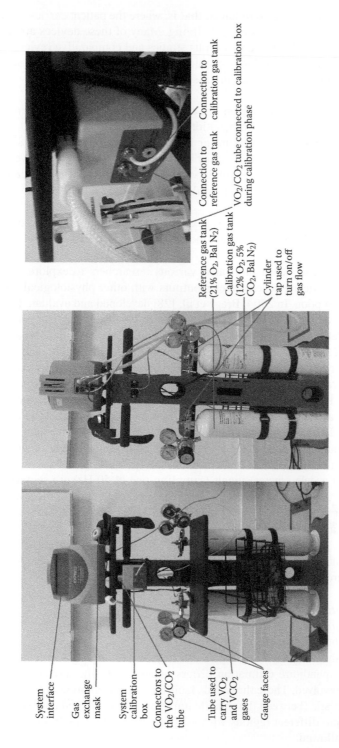

System interface

Gas exchange mask

System calibration box

Connectors to the VO₂/CO₂ tube

Tube used to carry VO₂ and VCO₂ gases

Gauge faces

Reference gas tank (21% O₂, Bal N₂)

Calibration gas tank (12% O₂, 5% CO₂, Bal N₂)

Cylinder tap used to turn on/off gas flow

Connection to reference gas tank

Connection to calibration gas tank

VO₂/CO₂ tube connected to calibration box during calibration phase

FIGURE 6.9 An indirect calorimeter.

at home or elsewhere in the community, that is, where the patient carries on with his or her usual ordinary activities of daily living. Many of these devices are available in the market and are described in various studies and comparative reviews [30,31]. Unfortunately, there are some biomechanical disadvantages impeding the sole use of accelerometers for energy expenditure assessment:

1. Torso-mounted accelerometers are unable to determine energy expenditure associated with isolated limb motion.
2. Accelerometers may give similar intrinsic outputs (resulting in similar energy expenditure estimates) for activities that nevertheless have quite differing PA levels, for example, walking level or inclined, ascending stairs, cycling or rowing, and walking unloaded or loaded with heavy objects.
3. The relationship between energy expenditure and accelerometer counts is linear and predictable at low and moderate physical activity levels only.

These shortcomings have motivated various researchers to explore alternative solutions, such as fusing accelerometer outputs with other physiological indicators of energy consumption. In 2004, Brage et al. [28] developed and evaluated a method (Branched Equation Model [BEM]) for measuring levels of energy expenditure by combining accelerometry with HR monitoring, demonstrating improved estimation of these parameters when tested in 12 normal male subjects. The approach relies on a set of rules, regression equations, and thresholds to estimate the energy expenditure. Thus, these parameters are estimated by means of the selected piecewise function (i.e., one out of the four available in the branched model) that best suits the level and intensity of the activity currently performed. This model is illustrated in Figure 6.10. The algorithm does not make a distinction between the type of movement or activity and assumes that the energy of the signals obtained from the accelerometer will be a "good-enough" feature for the estimation of energy expenditure. More specifically, the raw triaxial accelerometer signals (sampling rate of 50 Hz) are segmented into blocks, with a duration of 15 s each. Subsequently, each channel is processed using a band-pass filter in the region of 0.25 to 6 Hz (the upper limit of the filter bandwidth was chosen to attenuate high-frequency disturbances occurring when the swinging foot affects the ground during initial contact [32]), rectified, integrated, and finally aggregated to obtain a single value—accumulated accelerometer count (AAC). The algorithmic parameters are as follows:

1. *Input features*: AAC and HR.
2. *Conversion functions*: Converts the AAC and HR into calorie expenditure.
3. P_i: Weights for different conversion functions.
4. T_j: Threshold values. The relative efficacy of AAC and HR as features for energy expenditure estimation differs according to the level of physical activity involved. These thresholds facilitate the partitioning of the transfer characteristic (between the input features and the estimated energy expenditure) into different regions where different contributions from AAC and HR are allowed.

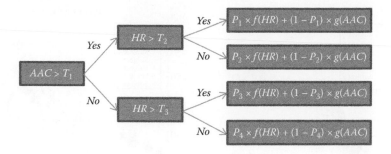

FIGURE 6.10   The BEM for calorie energy expenditure estimation.

More formally, the estimated calorie energy expenditure, $\hat{E}$, is defined in Equation 6.35.

$$\hat{E} = \begin{cases} P_1 f(HR) + (1 - P_1)g(AAC), & AAC > T_1, HR > T_2 \\ P_2 f(HR) + (1 - P_2)g(AAC), & AAC > T_1, HR \leq T_2 \\ P_3 f(HR) + (1 - P_3)g(AAC), & AAC \leq T_2, HR > T_3 \\ P_4 f(HR) + (1 - P_4)g(AAC), & AAC \leq T_2, HR \leq T_3 \end{cases} \qquad (6.35)$$

There are several variants of the BEM [28,33,34], which targets different populations and makes use of different sensors. As such, the algorithmic parameters are different. Therefore, it is imperative that the algorithmic parameters are optimized for our particular accelerometer and HR monitor. Having obtained the best estimates for HR and AAC, using a well-known HR algorithm [35] and the AAC algorithm described above, the objective is to estimate $f()$, $g()$, P_i, and T_j. To obtain the conversion functions, an experiment is designed in order to collect sensor data together with reference energy expenditure values, E, from an indirect calorimeter (Figure 6.9). This is carried out for eight healthy subjects, undergoing a variety of activities including walking/running on a treadmill, cycling on an ergometer, and stepping exercises. The data are aligned and processed in order to obtain the HR and AAC and correspondingly plotted in Figure 6.11. Note the breakpoints for both the HR and AAC transfer characteristics—these points demarcate regions of linearity in the transfer characteristics. Also, in the nonlinear regions, the variance is noticeably larger, indicating that better accuracy can be achieved by operating on the linear portion of the transfer functions. Therefore, piecewise regression is carried out in order to obtain the best approximation for the transfer characteristics. More specifically, second-order polynomials are used to approximate the nonlinear segments of the transfer characteristics.

The objective function, along with the constraints, can then be formulated as shown in Equation 6.36. In this case, the objective is to minimize the sum of absolute error across N data points between the reference value, E, and the corresponding predicted value, $\hat{E}$, which is represented by $\varepsilon_{T,P}$ (vectors $\mathbf{T}$ and $\mathbf{P}$ are populated by the current thresholds and weights being considered). The parameters AAC_k and HR_k

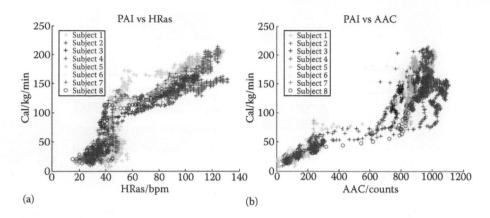

(a)   (b)

FIGURE 6.11   (a) HR and corresponding E. (b) AAC and corresponding E. The data set comprises data collected from an indirect calorimeter and corresponding HR and AAC values from eight subjects.

refer to the AAC and HR acquired from the sensors at instance k. The arguments T_i and P_j are to be adjusted to minimize $\varepsilon_{T,P}$. The parameter b refers to the resolution of the ADC of the accelerometer (which is 8 in our case). The parameters S and I refer to the sampling rate and the duration of one segment of data (which are 50 Hz and 15 s, respectively).

$$\underset{T_1,\dots,T_3,P_1,\dots,P_4}{\text{minimize}} \quad \varepsilon_{T,P} = \sum_{k=0}^{N-1} \left| E_k - \hat{E}(AAC_k, HR_k, T_1,\dots,T_3, P_1,\dots,P_4) \right|$$

$$\text{subject to} \quad \begin{aligned} &0 \leq T_1 \leq 3(2^{b-1}-1) \times S \times I \\ &0 \leq T_2, T_3 < 300 \\ &0 \leq P_i \leq 1, i = 1,\dots,4. \end{aligned} \tag{6.36}$$

It can be observed that the design space is rather large, consisting of seven parameters. In addition, the compute time for $\varepsilon_{T,P}$ varies with the number of data points, N, which is large. Therefore, an exhaustive search for the optimal solution is undesirable. On the other hand, heuristics-based search algorithms are more attractive options. Empirical observations indicate that $\varepsilon_{T,P}$ is a nonconvex function. For this reason, a nongreedy optimization algorithm, simulated annealing (SA), has been selected to search for a pseudo-optimum solution [36]. *SA* has been initially designed to control the annealing process of solids—gradual cooling of a substance from a liquid state to a solid state at the lowest level of energy. This algorithm has a few critical parameters, including the state of the algorithm, S (consisting of the tunable parameters), the number of points within the design space to search in the current iteration, L, and a control parameter, c. A large control parameter makes it less probable for the algorithm to accept the current state as the best solution so far, and vice versa. The algorithm has a number of phases [36]:

1. *Initialization*: The algorithm initializes the critical parameters, including S, L, and c.
2. *Initial exploration*: Note that each candidate solution (particular values for S) is generated randomly ("perturbation"). Better solutions are immediately accepted. Otherwise, poorer solutions may be accepted depending on c. This nonintuitive approach is carried out in order to prevent the algorithm from being stuck in a local minimum.
3. *Late exploration stages*: L and c are adjusted in each iteration. As *SA* converges, it becomes increasingly greedy and less likely to accept poor solutions.
4. *Stop criterion*: An arbitrarily low value for the optimization error function and a maximum number of iterations are the typically used stop criterions for *SA*.

To adapt our problem for *SA*, the state S refers to the set of weights and thresholds: $\{P_i, T_j | i = 1, \ldots, 4; j = 1, \ldots, 3\}$, and these are initialized according to parameters found in Ref. [28]. In addition, the error function, ε, is used to gauge the relative quality of solution candidates and as the stop criterion for the algorithm. Using this technique, an effective solution tuned to the sensor interface and other algorithmic parameters was located rapidly. The aggregate accuracy of the algorithm (both floating point and fixed point), in comparison with indirect calorimetry, can be found in Figure 6.12.

6.3.5 CASE STUDY 3: ARRHYTHMIA DETECTION (TYPE 2)

There are different types of cardiac aperiodic rhythms. AF is perhaps the most commonly seen in general wards and often leads to severe complications and life-threatening conditions when left untreated. Several experts have pointed out that AF is one of the most common types of arrhythmia in clinical practice [3,37,38]. The overall prevalence of this condition is between 1% and 1.5% in the general population. The prevalence increases with age, particularly in the elderly, that is, approximately 10% in those older than 70 years [15]. A report from the American Heart Association not only revealed an estimate of 2.66 million people with AF in 2010 but also predicted that such figure will increase by 12 million in 2050 in the United States [39].

This motivated different scientists and bioengineers to explore and develop effective methods for detecting AF that can be incorporated in clinical monitors. A large body of work has been carried out to distinguish between AF, normal rhythms, and other types of abnormal rhythms [37]. However, these methods tend to be suitable for patients who are at rest. The objective of this work is to present a three-way classifier that is capable of distinguishing ambulation noise from AF and normal rhythms, thus making it useful in clinical contexts such as the general ward, where patients are ambulatory, but could potentially deteriorate in a short span of time [40].

In the ECG, the electro-ionic activity of the normal heart is represented by the summation of different cell potentials taking place at different parts of the heart. This results in a number of signature waveforms or complexes representing the polarization and repolarization of different nerve cells across the organ. As shown in Figure 6.5a [14], the normal ECG exhibits a steady and regular rhythm. Deviation from these

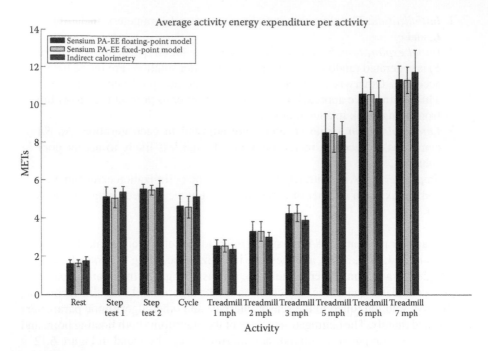

FIGURE 6.12 Bar chart of overall accuracy for floating-point and fixed-point versions of the calibrated BEM, in comparison with indirect calorimetry.

parameters may indicate the presence of an abnormal rhythm or heart condition. The characteristics of AF as reflected in the ECG (Figure 6.5b [15]) are as follows:

1. Very irregular R–R intervals (RRIs).
2. Absence of P waves.
3. Abnormal atrial activity (F waves—coarse or fine).
4. Although AF is often accompanied by rapid ventricular rates (>100 bpm), the frequency of ventricular contractions can vary between 50 and 250 bpm depending on the degree of AV conduction, patient age, and medications (such as beta-blockers) [41].

An extensive review of existing literature revealed that determining the extent of irregularity in RRIs will allow the accurate detection of AF. Indeed, extracting/selecting features from RRIs have yielded better performance than those based on hybrid methods or ECG morphological characteristics of AF. One of the most successful methods based on RRI is that proposed by Tateno and Glass [42]. This approach relies on the use of standard density histograms, and coefficients of variation of RRIs and their successive differences, in order to detect AF. The authors reported values of Se and Sp of 94.4% and 97.2%, respectively. In addition, Larburu et al. tested this algorithm and obtained classification performance values greater than 90% (see Table 6.1). Other researchers have also attempted to develop efficient classifiers using RRIs. Moody and Mark developed a method involving the use of

Markov's models to detect AF from RRI features. Results from the evaluation of this method using data from the MIT-BIH Arrhythmia Database showed *Se* > 90% but *Sp* < 90% and +*P* < 90%. Similar or poorer performances were observed for other RRI-based algorithms assessed by Larburu et al. (see Table 6.1).

As mentioned in Section 6.3.1, the types of data used for training the classification algorithm is of particular importance. In order to capture the impact of different types of artifacts on classification performance, we used a mixture of databases:

1. *Physiological artifacts*: the MIT-BIH Arrhythmia Database and the MIT-BIH AF Database [12]. In addition to AF, other atrial and ventricular arrhythmias including atrial flutter and atrial bigeminy are included in these databases. In addition, a small database of ECG signals was collected from patients using the target device during a clinical trial in a hospital. The patients are suffering from a number of comorbidities including high body mass index, diabetes, obesity, pulmonary edema, AF, congestive heart failure, and Wolff–Parkinson–White syndrome.
2. *System-generated artifacts*: ECG signals were collected from the laboratory and from the short clinical trial. System artifacts are implicitly included within the signals.
3. *Mechanical artifacts*: Some of the signals within the above databases exhibit substantial baseline wander, indicating movement by ambulating patients or the electrodes themselves.
4. *Prior distribution*: A MATLAB® Graphical User Interface (GUI)–based tool has been created for the selection of data and its randomization so that the prior distribution of the data can be controlled (Figure 6.13).

TABLE 6.1
Comparison between AF Detection Algorithms in Terms of Classification Performance/Accuracy

Algorithm	Se (%)	Sp (%)	PPV (%)	Err (%)
Moody and Mark [43]	87.54	95.14	92.29	7.88
Logan and Healey [44]	87.30	90.31	85.72	10.89
Linker [45]	97.64	85.55	81.81	9.61
Tateno and Glass [42]	91.20	96.08	90.32	5.32
Cerutti et al. [46]	96.10	81.55	75.76	16.62
Slocum et al. [47]	62.80	77.46	64.90	28.39
Novac et al. [48]	89.20	94.58	91.62	7.57
Babaeizadeh et al. [49]	87.27	95.47	92.75	7.80
Couceiro et al. [50]	96.58	82.66	78.76	11.77

Source: N. Larburu et al., Comparative study of algorithms for atrial fibrillation detection. In *Proc. IEEE Conference on Computing in Cardiology*, pages 265–268, Hangzhou, China, September, 2011.

Note: Err, classification error; PPV, positive predictivity; *Se*, sensitivity; *Sp*, specificity.

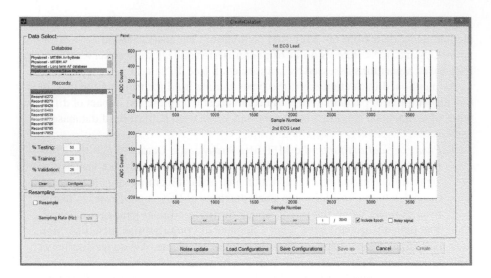

FIGURE 6.13 MATLAB GUI used for data selection and the specification of prior distribution (note that excluded data records for certain patients are highlighted in gray).

When the ECG signal is acquired by the front-end sensors, it is divided into segments of 30 s in duration. The RRIs are then extracted using an R-peak detection algorithm [35]. To enable more effective arrhythmia classification, each block of RRIs is preprocessed by applying a moving average (median) filter approach [51], as follows. First, this stage calculates the median RRI out of the first nine RRI windows, resulting in the first estimate for x_m. For new windows, x_m is updated by the mean of its current value and its previous value. A test statistic is then calculated as shown in Equation 6.37, where med{.} is the median operator applied over the current window and $x(n)$ corresponds to the RRI located in the middle of the window (i.e., fifth element in a window of nine elements, centered around index n). The denominator represents a robust approximation of the standard deviation of an equivalent Gaussian signal. The current RRI is replaced by the median interpolated estimate based on the criterion specified in Equation 6.38, and this is set to 4 based on McNames et al. [51].

$$D(n) = \frac{\left|x(n) - x_m\right|}{\mathrm{med}\left\{\left|x(n) - x_m\right|\right\}} \tag{6.37}$$

$$\hat{s}(n) = \begin{cases} x(n), & D(n) < \tau \\ \hat{s}(n), & D(n) \geq \tau \end{cases} \tag{6.38}$$

It is important to emphasize that the choice of this filter was made bearing in mind its effectiveness removing spurious information from ECG signals or reasonable quality, but without having a significant filter effect in signals fully corrupted by noise. Thus, the filter helps to improve the performance of the classifier when discriminating between different classes.

In order to understand the data distribution in the context of the feature space (defined by features selected by the user), and how certain types of data (particularly noise) can confound the classification process, the MATLAB GUI is extended for the purpose of feature space visualization (Figure 6.14a). A "probe" is used in order to investigate particular data points by double-clicking the point of interest—this facility allows the data point to be traced back to the particular database and the patient of origin (Figure 6.14b). This tool has two functions: first, it allows the features that would obtain the best separability between the target classes to be quickly identified; second, it allows data clusters that could potentially confound the classification results, particularly those on the boundaries of class separation, to be identified in terms of their data type (e.g., noise or the type of arrhythmia) and their location in feature space, and expose potential algorithmic limitations.

The visualization of the feature space is limited to three dimensions at a time. Automatic feature selection techniques that can be used beyond three dimensions, such as RELIEFF and mRMR, can be deployed for high-dimension feature selection. However, this algorithm is targeted for implementation on an embedded platform, where computing resource is scarce. In addition, the concurrent use of two

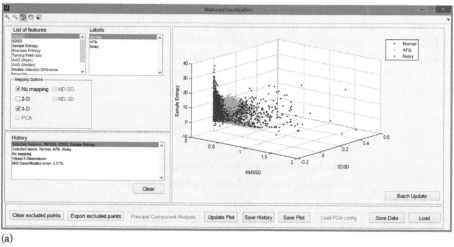

(a)

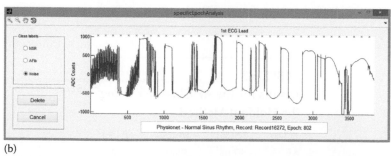

(b)

FIGURE 6.14 (a) MATLAB GUI used for feature space visualization. (b) Probe used to investigate the nature of the data point and trace it back to its point of origin.

features, RMSSD and HR (mean NN interval), resulted in an accurate classifier. Consequently, automatic feature selection techniques were not used in this design context. After the exploration of different classification options (i.e., neural networks, SVMs, and discriminant analysis) using two candidate feature spaces, we found that three-way classifiers based on decision trees met the best compromise between classification performance and the low computational cost.

Among the data chosen, the start and end times of each AF episode are recorded. However, some of them are too short to be meaningfully detected. For that reason, an episode is considered to contain an AFib rhythm only if it contains 15 s of the rhythm or longer. Other characteristics of the database are as follows. Table 6.2 shows the relative sizes of the randomly created partitions used in training and testing the AF classification algorithm.

1. The vast majority of the signals corresponded to normal rhythms.
2. A large proportion of the AF signals encompass rapid rhythms between 120 and 190 bpm.
3. The proportion of noise signals is considerably smaller in comparison to the other two classes.

The tree classifier used in the determination of AF rhythms is derived, as shown in Figure 6.3. The holdout method is described in Section 6.3.2, with the test set being used for the accuracy evaluation of the AF classifier. Since this problem consists of three classes, AF, NSR, and noise, the classification results for each class (defined as the +Class) is shown in Table 6.3.

TABLE 6.2
Data Set Partitions for Training and Evaluating the AF Classification Algorithm

Partition	Size	Proportion (%)
All	2246	100
Training	1568	69.8
Testing	678	30.2

TABLE 6.3
Results for the AF Classifier Using the Holdout Method

+Class	*Se* (%)	*Sp* (%)	+*P* (%)
AF	97.04	96.08	94.24
NSR	98.11	99.76	99.62
Noise	90.21	98.13	92.81

6.3.6 CASE STUDY 4: VITAL SIGNS PREDICTION (TYPE 3)

In an automatic early warning system, the clinical staff are notified if their charges are exhibiting abnormal physiological vital signs. Certain vital signs, such as the respiratory rate, are particularly susceptible to artifacts arising from motion and other sources. These artifacts confound the notification system and potentially desensitize the attending nurse from actual clinical events, when immediate medical attention is actually required. Therefore, one option is to apply effective filters to attenuate these artifacts in order to obtain reliable vital signs. The second option is to determine the reliability of the derived vital sign and report an error code instead if the vital sign is deemed to be unreliable. Under tight computational and energy constraints, the second option is usually more viable than the first. This is particularly true in telemetry systems when the raw physiological signal is acquired and processed on-chip to obtain the vital signs, before it is wirelessly streamed to a server for storage and display [52]. Such a scenario might give rise to a high frequency of error codes from ambulating patients, which is unacceptable for the monitoring staff.

One solution would be to use a vital signs predictor, based on historical trends. There are two purposes for such a predictor. First, the reliability of the current prediction could be more accurately ascertained as vital signs tend to vary slowly across time. Second, the predicted vital sign could be reported in place of the error code. Clearly, there is a danger that the predicted vital sign might not be representative at that point in time. However, this problem can be resolved by waiting for a requisite amount of time for the predicted value and the actually derived vital sign to converge. If convergence does not take place, an error code can then be reported and a notification can be raised.

The above scenario can be modeled as a linear dynamic system, where the underlying signal is corrupted by additive white noise. In this case, the underlying signal consists of vital signs derived from the patient at discrete time instances—this signal is denoted by z_n, which refers to the true vital sign of a patient at time instance n. This signal is corrupted by sources, which could include the colored noise from the hardware as well as motion. It is assumed that, when aggregated, the noise can be modeled as a zero-mean Gaussian distributed random variable. It is further assumed that the observed sequence, x_n, is a linear combination of noise and the underlying signal. Therefore, since z_i is unknown, the objective is to estimate its value at each time instance, given the observed variable x_n. Since the vital sign of a patient at one time instance is likely to be similar to that in the next time instance, the vital sign detected at a time instance, z_n, can be modeled as a linear function of the vital sign from the previous time instance, z_{n-1}, as shown in Equation 6.39. The errors from this model are normally distributed around zero, with a variance of Γ. Similarly, the actual measurement, x_n, is linearly related to the state, and expressed in Equation 6.40, with additive zero-mean Gaussian noise. Equations 6.39 and 6.40 are known as the Kalman filter equations [11].

$$z_n = Az_{n-1} + w_n \tag{6.39}$$

$$x_n = Cz_n + v_n \tag{6.40}$$

$$w \sim N(0, \Gamma) \tag{6.41}$$

$$v \sim N(0, \Sigma) \tag{6.42}$$

The Kalman filter equations share remarkable similarities with the HMM, which is described in Section 6.2.3. Indeed, the state and observed variables, z_n and x_n, are akin to a latent variable C_n and feature F_n in Figure 6.4. Both systems attempt to capture the correlation between consecutive states, as shown in Equations 6.17 and 6.39. The difference between the HMM and the Kalman filter is in the definition of the posterior marginal probability, since the state variable is discrete in the case of the HMM and continuous in the case of the Kalman filter. Specifically, the posterior marginal probability, $\alpha(z_n)$, is given by Equation 6.43, which has a form similar to Equation 6.24 [7].

$$\alpha(z_n) = P(x_n | z_n) \int \alpha(z_{n-1}) P(z_n | z_{n-1}) \, dz_{n-1} \qquad (6.43)$$

Assuming that the current state is given by z_i at time instance i, the a priori prediction for the next state can be computed by taking $\hat{z}_{i+1}^- = A z_i$. Since the state variable, z_i, is a noisy version of observation x_i, they are closely related. It follows that prediction errors for both variables are correlated. Therefore, a better estimate, or the a posteriori estimate (denoted by $\hat{z}_i$), can be obtained using Equation 6.44. The gain parameter K_i is adapted in each time step in order to minimize the covariance metric, P_i.

$$\hat{z}_i = \hat{z}_i^- + K_i(x_i - C z_i) \qquad (6.44)$$

$$P_i = E\{(z_i - \hat{z}_i)(z_i - \hat{z}_i)^T\} \qquad (6.45)$$

In fact, from Ref. [11], P_i is minimized if the gain parameter takes the form shown in Equation 6.46. The corresponding covariance, P_i, is shown in Equation 6.47.

$$K_i = P_i^- C_i^T \left[C_i P_i^- C_i^T + R_i \right]^{-1} \qquad (6.46)$$

$$P_i = (I - K_i C_i) P_i^- \qquad (6.47)$$

In order for the Kalman filter to be deployed, its parameters have to be set appropriately. This is best explained within the application context. In this case, we are using the Kalman filter as a predictor of respiration rate for hospital-bound patients, in order to attenuate motion artifact noise and provide vital signs to clinical staff (on the basis of past measurements) when the current data segment is deemed to be too corrupted for a reliable rate to be produced. In this case, we assume that $\mathbf{z} = \{z_0, z_1, ..., z_4\}$—five previous states were used in the prediction of the current state. This approach was found to be capable of predicting the respiration rate compared to the use of only the most recent state. The respiration rate is calculated using an algorithm [53], which preprocesses the signal in order to determine if the resulting vital sign is reliable. If it is deemed reliable, it is assigned to measurement x_i. Otherwise, x_i

is assigned to the most recent and reliable measurement available. More specifically, the parameters of the filter are determined as follows:

1. Transition matrix A and transfer matrix C: this is determined using regression (minimization of the least square), as shown in Equations 6.48 and 6.49, respectively [54].
2. State vector $\mathbf{x}_0$: this is initialized with the first five detected respiration rates.
3. Covariances $\mathbf{w}$ and $\mathbf{v}$: these parameters are derived empirically from training data.

Note that some of the parameters are derived from a training set, containing data collected from actual patients (to enable rapid convergence).

$$A = \arg\min_{A} \sum_{k}^{M-1} \left\| \mathbf{z_{k+1}} - A\mathbf{z_k} \right\|^2 \tag{6.48}$$

$$C = \arg\min_{C} \sum_{k}^{M} \left\| x_k - Cx_k \right\|^2 \tag{6.49}$$

During the initialization phase of the filter, the Kalman filter is not invoked until the state vector, $\mathbf{z}$, has been fully populated. Subsequently, the Kalman filter is invoked by carrying out two steps. The first step involves the a priori estimate of the state and the error covariance matrix using Equations 6.50 and 6.51. The second step involves the a posteriori estimate of the state and the covariance, as seen in Equations 6.52 through 6.54. The aggregated respiration output, $\mathbf{z}_i^{\text{agg}}$, is derived by taking the median of the a posteriori state vector, as shown in Equation 6.55. This implementation results in a real-time system, where both steps take place cyclically, producing one filtered respiration rate for every raw input rate.

$$\hat{\mathbf{z}}_\mathbf{i}^- = A\hat{\mathbf{z}}_{i-1} \tag{6.50}$$

$$P_i^- = AP_{i-1}A^T + w \tag{6.51}$$

$$K_i = P_i^- C^T \left[CP_i^- C^T + R \right]^{-1} \tag{6.52}$$

$$\hat{\mathbf{z}}_\mathbf{i} = \hat{\mathbf{z}}_\mathbf{i}^- + K_i \left(x_i - C\hat{\mathbf{z}}_\mathbf{i}^- \right) \tag{6.53}$$

$$P_i = (I - K_i C)P_i^- \tag{6.54}$$

$$\mathbf{z}_\mathbf{i}^{\text{agg}} = \text{median}(\hat{\mathbf{z}}_\mathbf{i}) \tag{6.55}$$

A sample of the Kalman filtered outputs, along with the raw input respiration rates, is shown in Figure 6.15. Two observations can be made: first, local fluctuations in the respiration rates have been reduced as the number of spikes within the trend has been reduced; second, between the 120th and the 130th segments, an upward trend in respiration rate is observed and the Kalman filter was able to track this trend rapidly.

In aggregate, the distribution of the prediction errors is illustrated in Figure 6.16. This is derived from the histogram composed of the differences between the current measurement (x_i) and the "predicted measurement" $\left(C\hat{z}_i^-\right)$, for each invocation of the filter (part of Equation 6.53). Clearly, this distribution bears strong visual resemblance to a Laplace distribution [55]. The mean value for the errors is 0.27 brpm and the 95% prediction interval is [−5.34, 5.88] brpm.

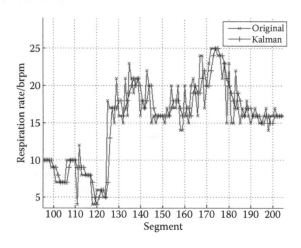

FIGURE 6.15   Inputs x_i and outputs $\left(z_i^{agg}\right)$ of the Kalman filter.

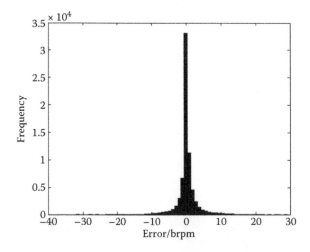

FIGURE 6.16   Distribution of prediction errors: $x_i - C\hat{z}_i^-$, $N = 71{,}798$.

The evaluation of the filter described above is by no means comprehensive. The difficultly of doing so arises from the fact that no reference respiration rates from third-party devices exist. If reference data are available, correlation analysis and statistical significance tests (such as Student t and Wilcoxon signed rank tests [13]) could be carried out to ascertain the accuracy of the respiration rates derived from the filter.

6.4   DISCUSSION

The main objective of this chapter is to create a framework, allowing problems and their constraints to be formulated clearly. Once the problem has been properly distilled, the metrics for evaluation defined, and the corresponding set of candidate solutions or algorithms identified, how should one go about implementing the solution? A design and build flow, which we found useful, is shown in Figure 6.17.

The exploration of new or existing approaches involves a comprehensive review to identify existing algorithms that can be used to solve the formulated problem in the light of existing constraints. Some of the approaches may be unsuitable because of these constraints. For example, a real-time solution might be required to solve the problem at hand, implying that future data are not available in order to produce a result at each time instance. This might rule out approaches such as the HMM as a means of solving the problem. Once the potential algorithm candidate set has been determined, they are prototyped in flexible programming languages such as MATLAB [56], which allows rapid implementation and comprehensive analysis of their relative performance. Further, a representative data set needs to be collected. The guidelines for its collection and evaluation are detailed in Section 6.3.1. In the case of classification problems, training methods for determining the classifier parameters are discussed in Section 6.3.2. Candidates that do not meet performance requirements are culled from the solution set at this point.

The most promising candidates are then ported over to a programming language compatible with the target device. An example is C, which is the language of choice for many embedded processors. The programming constructs used in this case are kept as generic as possible so that the algorithm candidates are portable across multiple platforms. Subsequently, an initial algorithm profile is carried out to determine the resource requirements, including the code, data memory space requirements, and its energy consumption. Algorithm candidates that do not meet these platform constraints are removed. At this stage, an additional data collection stage might be carried out to further test the robustness of the algorithm. For instance, in the case of an HR computation algorithm, ECG signals might be collected from ambulating patients who are known to have a variety of arrhythmias and subsequently applied to the candidate algorithms. If the algorithmic performance or its computational requirements are deemed to be too expensive, further optimization may be necessary to ensure that the platform or performance constraints are met. This is often an iterative process, which would carry on until these constraints are met.

Downstream of the necessary checks for regulatory compliance, the selected algorithm has to be integrated into the target embedded platform. At this stage, the Application Programmer Interface (API) functions for the algorithm, test vectors,

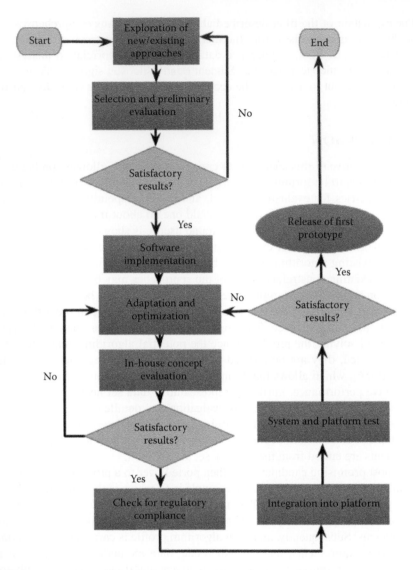

FIGURE 6.17   Design and build flowchart for biomedical algorithms.

and corresponding results should be made available. Having these API functions allow the algorithms to be easily deployed without the need for changing the underlying source code. The test vectors and their expected results provide the means of checking if the functionality of the algorithm has been preserved during the integration process. As an example, double precision floating-point arithmetic operations that are common in MATLAB are implicitly reduced to single precision in the Keil environment [57] (programming environment for several embedded processors including the Intel 8051 eWarp processor [52]). In applications such as high-order IIR filters, this might create unacceptable differences in algorithmic behavior. Therefore, the integration data set should be large and representative enough in order to expose these differences and any algorithmic limitations.

An overall system test is then carried out to determine if the system works well with the newly integrated algorithm. System-level bugs might be revealed at this stage. For instance, intermediate buffers for dynamic data storage might not be properly allocated and removed within the algorithm during run-time, potentially causing memory corruption in other modules, such as the software radio module responsible for the wireless transmission of the payload. Therefore, a comprehensive set of test cases should be defined to represent realistic scenarios in which the system might be used, so as to expose system-level bugs, which can then be corrected. Finally, after the system has passed the battery of tests, the first prototype can be released for deployment in the field.

6.5 CONCLUSION

On the basis of our observations, we found that problems encountered in the healthcare monitoring industry can each be categorized into one of three types: generic optimization problem, static classification problem, and prediction problem. There are substantial overlaps between these problems, and algorithms are often suited for multiple problems. At times, the problem type is made on the basis of certain application constraints. For instance, by formulating a problem as one of a static classification task, rather than a dynamic predictor, one is able to create a more compact and efficient classifier that can be realized on the target platform.

In addition, the data set used in the training and evaluation of the algorithm is crucial, as it affects the algorithmic parameters, the relative results between algorithm candidates, and potentially the choice of algorithm. A representative and comprehensive data set is often difficult to come by, and relevant data are often dependent on the sensor front-end used for data collection as well as the physiological condition of the patient. In order to solve this problem, guidelines for initial data collection in controlled laboratory conditions, together with methods to augment the data set, are presented in this chapter. This is particularly important for static classification problems. For predictive problems, the initial data set and algorithmic parameters are important because the predictive outcomes of such algorithms are dependent on the prior data distribution during early runs of the algorithm and they may take a longer time to converge on the correct results should the initial data set be of bad quality.

In order to demonstrate the formulation framework, and the thought processes involved during algorithm design, four case studies involving different types of problems were provided. Note that in each case, only a cursory discussion on the statistical evaluation of the results was given because they are outside the scope of this chapter. Apart from the algorithm design, its implementation aspects were also discussed in Section 6.4, with emphasis on algorithmic portability across different platforms and a robust testing framework—from the initial unit testing, integration testing, to the final system testing.

Consequently, it can be concluded that in conjunction with a representative data set, problem formulation is an especially critical part of the design process. They provide a means by which problems can be described concisely, as well as a framework within which to automate problem solving. The authors acknowledge that there may be problems that cannot be categorized into any of the three types described, but this work is the result of actual problems encountered in the field and our insights and attempts at solving them.

REFERENCES

1. R. S. Garfinkel and G. L. Nemhauser. 1970. *Integer Programming*. John Wiley & Sons, UK.
2. V. Kumar P. N. Tan, and M. Steinbach. 2005. *Introduction to Data Mining*. Addison-Wesley, Boston.
3. L. Sornmo, M. Stridh, D. Husser, A. Bollmann, and S. B. Olsson. 2009. Analysis of atrial fibrillation: From electrocardiogram signal processing to clinical management. *Philosophical Transactions of the Royal Soc. A*, (28):235–253.
4. C. Cortes and V. Vapnik. 1995. Support-vector networks. *Machine Learning*, (3):273–297.
5. F. Long, H. C. Peng, and C. Ding. 2005. Feature selection based on mutual information: Criteria of max-dependency, max-relevance, and min-redundancy. *IEEE Transactions on Pattern Analysis and Machine Intelligence*, (8):1226–1238.
6. M. Robnik-Šikonja and I. Kononenko. 2003. Theoretical and empirical analysis of reliefF and RReliefF. *Machine Learning*, (53):23–69.
7. C. M. Bishop. 2006. *Pattern Recognition and Machine Learning*. Springer, Cambridge, UK.
8. J. W. Sammon. 1969. A nonlinear mapping for data structure analysis. *IEEE Transactions on Computers*, (18):401–402.
9. J. C. Principe, C. Lefebvre, and C. L. Fancourt. 2013. Artificial neural networks. In P. M. Pardalos and H. E. Romejin, editors, *Handbook of Global Optimization*, volume 2, pages 363–386.
10. J. R. Quinlan. 1993. *C4.5: Programs for Machine Learning*. Morgan Kaufmann Publishers, USA.
11. L. Mason. 2002. *Signal Processing Techniques for Non-Invasive Respiration Monitoring*. PhD thesis, Oxford University, UK, October.
12. A. L. Goldberger, L. A. N. Amaral, L. Glass, J. M. Hausdorff, P. Ch. Ivanov, R. G. Mark, J. E. Mietus, G. B. Moody, C.-K. Peng, and H. E. Stanley. 2000. PhysioBank, PhysioToolkit, and PhysioNet: Components of a new research resource for complex physiologic signals. *Circulation*, (23):e215–e220.
13. S. Navas. 2003. Atrial fibrillation: Part 1. *Nursing Standard*, (17):45–54.
14. S. Goodacre and R. Irons. 2002. ABC of clinical electrocardiography: Atrial arrhythmias. *British Medical Journal*, (324):594–597.
15. C. Redmond and T. Colton. 2001. *Clinical Significance versus Statistical Significance*. John Wiley & Sons, UK.
16. R. Kohavi. 1995. A study of cross-validation and bootstrap for accuracy estimation and model selection. In *International Joint Conference on Artificial Intelligence*, pages 1137–1143, Canada, August.
17. W. D. Penny, D. Husmeier, and S. J. Roberts. 1999. An empirical evaluation of Bayesian sampling with hybrid Monte Carlo for training neural network classifiers. *Neural Networks*, (12):677–705.
18. N. Metropolis, A. W. Rosenbluth, M. N. Rosenbluth, A. H. Teller, and E. Teller. 1953. Equation of state calculations by fast computing machines. *Journal of Chemical Physics*, (21):1087–1092.
19. G. K. Wallace. 1991. The JPEG Still Picture Compression Standard. *Communications of the ACM*, (34):30–44.

20. E. U. K. Melcher, L. V. Batista, and L. C. Carvalho. 2001. Compression of ECG signals by optimized quantization of discrete cosine transform coefficients. *Journal of Medical Engineering and Physics*, (23):127–134.

21. R. M. Neal, I. H. Witten, and J. G. Cleary. 1987. Arithmetic coding for data compression. *Communication ACM*, (6):520–540.

22. J. A. Nelder and R. Mead. 1965. A simplex method for function minimization. *The Computer Journal*, (4):308–313.

23. F. B. Hu. 2003. Sedentary lifestyle and risk of obesity and type 2 diabetes. *Lipids*, (2): 103–108.

24. A. Alwan. 2010. Global Status Report on Non Communicable Diseases. Available at http://apps.who.int/iris/bitstream/10665/44579/1/9789240686458_eng.pdf. Accessed July 2015.

25. J. Weir. 1949. New methods for calculating metabolic rate with special reference to protein metabolism. *Journal of Physiology*, (109):1–9.

26. D. C. Simonson and R. A. Defronzo. 1990. Indirect calorimetry: Methodological and interpretative problems. *American Journal of Physiology-Endocrinology and Metabolism*, (258):E399–E412.

27. K. R. Westrup and G. Plasgui. 2007. Physical activity assessment with accelerometers: An assessment against doubly-labelled water. *Journal of Obesity*, (15):2371–2379.

28. N. Brage, S. Brage, and P. W. Franks. 2004. Branched equation model of simultaneous accelerometry and heart rate monitoring improves estimate of directly measured physical activity energy expenditure. *Journal of Applied Physiology*, (96):343–351.

29. H. P. Johansson, L. Rossander-Hulthén, F. Slinde, and B. Ekblom. 2006. Accelerometry combined with heart rate telemetry in the assessment of total energy expenditure. *Journal of Nutrition*, (95):631–639.

30. S. Murphy. 2009. Review of physical activity measurement using accelerometers in older adults: Considerations for research design and conduct. *Journal of Preventive Medicine*, (48):108–114.

31. D. Andre and D. Wolf. 2007. Recent advances in free-living physical activity monitoring: A review. *Journal of Diabetes Science and Technology*, (1):760–767.

32. E. Antonsson and R. Mann. 1985. The frequency content of gait. *Journal of Biomechanics*, (18):39–47.

33. J. Churilla, S. Crouter, and D. Bassett. 2007. Accuracy of the actiheart for the assessment of energy expenditure in adults. *European Journal of Clinical Nutrition*, (62):1–8.

34. J. McClain and C. T.-Locke. 2009. Objective monitoring of physical activity in children: Considerations for instrument selection. *Journal of Science and Medicine in Sports*, (12):526–533.

35. J. Pan and W. J. Tompkins. 1985. A real-time QRS detection algorithm. *IEEE Transactions of Biomedical Engineering*, (3):230–236.

36. J. Korst, E. Aarts, and W. Michiels. 2005. *Search Methodologies*. Springer, USA.

37. N. Larburu, T. Lopetegi, and I. Romero. 2011. Comparative study of algorithms for atrial fibrillation detection. In *Proc. IEEE Conference on Computing in Cardiology*, pages 265–268, Hangzhou, China, September.

38. F. Yaghouby, A. Ayatollahi, R. Bahramali, M. Yaghouby, and A. H. Alavi. 2010. Towards automatic detection of atrial fibrillation: A hybrid computational approach. *Computers in Biology and Medicine*, (40):919–930.

39. D. Lloyd-Jones, R. J. Adams, T. M. Brown, M. Carnethon, S. Dai, G. De Simone, T. B. Ferguson, E. Ford, K. Furie, C. Gillespie, A. Go, K. Greenlund, N. Haase, S. Hailpern, P. M. Ho, V. Howard, B. Kissela, S. Kittner, D. Lackland, L. Lisabeth, A. Marelli, M. M. McDermott, J. Meigs, D. Mozaffarian, M. Mussolino, G. Nichol, V. L. Roger, W. Rosamond, R. Sacco, P. Sorlie, R. Stafford, T. Thom, S. Wasserthiel-Smoller, N. D. Wong, J. Wylie-Rosett, and on behalf of the American Heart Association Statistics

Committee and Stroke Statistics Subcommittee. 2009. Heart disease and stroke statistics—2010 update: A report from the American Heart Association. *Circulation*, (121):e46–e215.

40. S. Hugueny, D. A. Clifton, and L. Tarassenko. 2011. Probabilistic patient monitoring with multivariate multimodal extreme value theory. *Biomedical Engineering Systems and Technologies*, (127):199–211.

41. S. Kara and M. Okandan. 2007. Atrial fibrillation with artificial neural networks. *Pattern Recognition*, (40):2967–2973.

42. K. Tateno and L. Glass. 2001. Automatic detection of atrial fibrillation using the coefficient of variation and density histograms of RR and delta RR intervals. *Medical and Biological Engineering and Computing*, (39):664–671.

43. G. B. Moody and R. G. Mark. 1983. A new method for detecting atrial fibrillation using R–R intervals. *Computers in Cardiology*, (10):227–230.

44. B. Logan and J. Healey. 2005. Robust detection of atrial fibrillation for a long term telemonitoring system. *Computers in Cardiology*, (32):619–622.

45. D. T. Linker. 2006. Long-term monitoring for detection of atrial fibrillation. US Patent 20060084883. Hughes Aircraft Company, Los Angeles, CA, April.

46. S. Cerutti, L. T. Mainardi, A. Porta, A. M. Bianchi. 1997. Analysis of the dynamics of RR interval series for the detection of atrial fibrillation episodes. *Computers in Cardiology*, (24):77–80.

47. J. Slocum, A. Sahakian, and S. Swiryn. 1992. Diagnosis of atrial fibrillation from surface electrocardiograms based on computer-detected atrial activity. *Journal of Electrocardiology*, (25):1–8.

48. D. Novac, M. Pekhun, R. Schmidt, and M. Harris. 2008. Atrial fibrillation detection. Patent Cooperation Treaty. Google Patents.

49. S. Babaeizadeh, R. E. Gregg, E. D. Helfenbein, J. M. Lindauer, and S. H. Zhou. 2009. Improvements in atrial fibrillation detection for real-time monitoring. *Journal of Electrocardiology*, (42):522–526.

50. R. Couceiro, P. Carvalho, J. Henriques, M. Antunes, M. Harris, and J. Habetha. 2008. Detection of atrial fibrillation using model-based ECG analysis. In *International Conference on Pattern Recognition*, pages 1–5, Florida, USA, December.

51. J. McNames, T. Thong, and M. Aboy. 2004. Impulse rejection filter for artifact removal in spectral analysis of biomedical signals. In *Proceedings of the IEEE Medical Biology Society*, pages 145–148, Haifa, Israel, September.

52. A. C. W. Wong, D. McDonagh, O. Omeni, C. Nunn, M. Harris, M. Silveira, and A. J. Burdett. 2009. Sensium: An ultra-low-power wireless body sensor network platform: Design and application challenges. In *Engineering in Medicine and Biology*, pages 6576–6579, Minneapolis, USA, September.

53. S.-S. Ang, M. A. Hernandez-Silveira, and A. Burdett. 2014. Challenges and trade-offs involved in designing embedded algorithms for a low-power wearable wireless monitor. In *15th International Conference on Biomedical Engineering*, pages 416–419, Singapore, December.

54. Y. Gao, E. Bienenstock, M. Serruya, W. Wu, M. J. Black, and J. P. Donoghue. 2002. Inferring hand motion from multi-cell recordings in motor cortex using a kalman filter. In *SAB Workshop on Motor Control in Humans and Robots: On the Interplay of Real Brains and Artificial Devices*, pages 66–73, Edinburgh, August.

55. K. Taesu, T. Eltoft, and T.-W. Lee. 2006. On the multivariate Laplace distribution. *Signal Processing Letters, IEEE*, (5):300–303.

56. MATLAB. The MathWorks, Inc.

57. ARM holdings. Keil compiler, tools by ARM. Available at http://www.keil.com.

7 Cooperative Data Fusion for Advanced Monitoring and Assessment in Healthcare Infrastructures

Vasileios Tsoutsouras, Sotirios Xydis, and Dimitrios Soudris

CONTENTS

7.1 INTRODUCTION

Nowadays, electronic equipment and sensor technology have matured enough to enable the widespread use of medical sensors for the constant monitoring of various biological aspects of a patient. Consequently, the current trend in the design of medical systems is to incorporate an ever-increasing number of sensors in the complete system architecture. Manufacturing and incorporating reliable and accurate sensors in a medical system is of great importance, and at the same time, the interpretation of sensor data and knowledge extraction is of the same criticality. In wearable systems, sensor data are combined and correlated to produce a human interpretable outcome. We refer to this procedure as data fusion being executed in a computational machine by a data fusion engine.

Because of the importance of extracting knowledge from incoming sensor data, there has been extensive research work conducted to formulate algorithms that correlate them both in a fast and efficient way. A major leap toward this direction was the open distribution of medical data from numerous medical-related databases worldwide, on the basis of which researchers were able to incorporate machine-learning techniques into their sensor data fusion engine designs. A distinctive example is the MIT-BIH ECG and EEG signal database [1], which has been used as a starting point to utilize sophisticated classification algorithms for the design of medical systems and eventually create the necessary software and hardware infrastructure to incorporate them into portable computational devices [2].

However, as the number of incorporated sensors increases, it is either not computationally feasible or not efficient enough to design a system around a stand-alone medical device. Consequently, a state-of-the-art system design includes more than one information flow regarding the sensor data, and different devices have access to a subset of these flows. On these devices, instantiated data fusion engines are different and they cooperate in order to successfully produce the final fusion algorithm. This chapter introduces a generalized data fusion engine scheme that can be configured to meet the data fusion requirements in a complete medical-related sensor-based system. More specifically, in Section 7.3 we describe a general system architecture supporting cooperative data fusion. Sections 7.4 and 7.5 explain the nature of the generalized data fusion scheme employed in our study while Section 7.6 explores cooperative data fusion through a real-life use case of an infrastructure supporting sensor deployment for wearable devices targeting wound monitoring and management. Section 7.7 presents in more detail the experimental setup and the evaluation/ validation of the proposed data fusion, while Section 7.8 concludes the basic points introduced in this chapter.

7.2 SYSTEM ARCHITECTURE FOR COOPERATIVE DATA FUSION

Figure 7.1 depicts an overview of the information flow of medical data within a sensor-based system. At the lowest level of hierarchy, there are the sensors, which form the main data sources of the system. One level upward, one finds devices located on site of the user, that is, the sensor hub and the wearable medical device, and on the final level, there is the clinical back-end usually referring to the server infrastructure of the medical site.

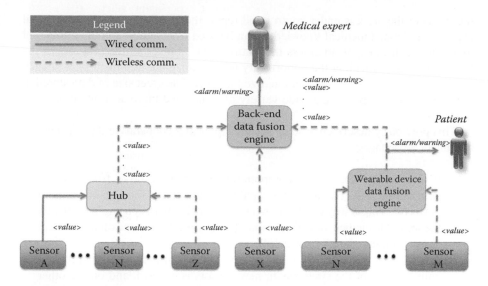

FIGURE 7.1   Information flow from data sources to data fusion engines.

As depicted in Figure 7.1, there are two data fusion engines that cooperate with each other, that is, the one located at the wearable device and the one located at the back-end data fusion system. The data fusion engine at the wearable device collects data from the integrated wearable sensors. The wearable device acts as both a localized monitoring infrastructure with immediate access to the patient and a proxy to transfer the medical data to the clinical back-end server. Its data fusion engine extracts information from the sensor readings and, whenever a warning or alarm is recognized, activates the transmission of a warning message to the patient while at the same time transmits the warning/alarm along with the respective sensor readings to the clinical back-end.

The data fusion engine at the wearable device is used in the embedded system as the basis to control the device. Therefore, the actual extracted information is directly used. Such controls are performed within the device based on predefined decision structures. One example for the functions controlled is the display of information on a graphical user interface that instructs the patient using the device to contact the treating physician when a sensor reading is out of its normal bounds. Hence, the function of the wearable device is adapted considering information about technical device parameters and aspects of the health status parameters of the patient, respectively.

Sensor data that are not forwarded to the wearable devices are transmitted to the clinical back-end either directly in a wireless manner as sensor X or through a hub device that aggregates data from a local sensor either wired or wirelessly and then it transmits the collected data to the clinical back-end. There, the second data fusion engine is allocated, where all the available information is aggregated. As expected, the data fusion engine located at the clinical back-end server is responsible for analyzing larger volumes of currently available and history data in order to assess the status of the process under control and report these assessments to the medical experts. Predefined medical knowledge can be used as an automatic diagnosis tool to aid the medical expert to assess the status of the wound and the progress of the healing process, providing the opportunity to customize the treatment plan of each patient in a more proactive and personalized manner. The great size of data stored in the server will help produce a wide variety of results and more accurate prediction of trends.

This previously described system design supporting cooperative data fusion has two major advantages:

1. It promotes off-loading of computationally intensive tasks from devices with diminished computational capabilities to devices that are much more powerful. This is why there are two instantiated data fusion engines, one on the wearable medical device (computationally weak) and one on the clinical server (computationally powerful).
2. The nature of the system is inherently modular. Thus, on the one hand, easy enhancement of the system with new sensors is enabled, and on the other hand, the system can function with different subsets of sensors tailored to the requirements of a specific user while minimizing the cost of the entire product.

7.3 A GENERAL MODEL FOR DATA FUSION ARCHITECTURE

The combination and interpretation of sensor derived data is mostly referred to as the sensor fusion task. Especially for wearable devices that interact with human-related activities, the importance of the task renders mandatory its correct function in a fast and effective way. For this to be realized, we show a modular design approach for the data fusion framework, where each module is bound to a specific type of data management. We refer to these modules as levels mainly because they can be abstractly perceived as ascending in sophistication and correlation of input data.

We designed a general and modular data fusion scheme. Generality enables the same data fusion scheme to be utilized both at the wearable medical device and the clinical back-end fusion engines. Modularity enables the data fusion framework to be tailored to differing system configurations, that is, configurations related to different scenarios regarding sensor availability and volume and maturity of data available during the evolution of a product. In addition, modularity allows the data fusion engines to be updated accordingly to the needs of the project with no need of reforming the main components of the software or hardware design.

Figure 7.2 shows an overview of the data fusion scheme adopted to implement the skeleton of data fusion engines both on the wearable device and on the server located at the clinical site. The execution model is formed as follows: after acquiring and storing the sensor readings (different sensor readings according to the location of the data fusion engine, i.e., wearable device, clinical back-end), the data fusion framework is executed. As depicted, the fusion algorithm can generate either alarms related to the detection of mechanical malfunctions or warnings related to the

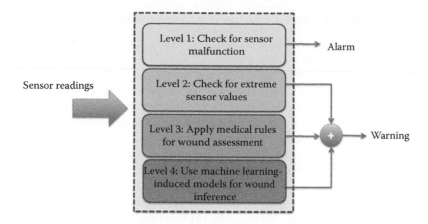

Sensor readings

Level 1: Check for sensor malfunction → Alarm

Level 2: Check for extreme sensor values

Level 3: Apply medical rules for wound assessment

Level 4: Use machine learning-induced models for wound inference

+ → Warning

FIGURE 7.2   The general scheme of data fusion framework.

detection of medical-related critical situations. Four levels of increasing algorithmic complexity are defined, each one implementing processing or filtering functions of a specific type.

7.3.1   Level 1: Alarm Generation for Sensor and Pump Malfunction

This is the first and most elementary level of sensor reading interpretation. In this level, sensor malfunction is recognized by the controlling processor and action is taken to notify the appropriate medical and technical experts and above all to ensure that there is no risk to the patient's health. We refer to the outcome of this level as an alarm in contrast to the following levels of fusion where we refer to warning generation. As a consequence and since the alarms are directly connected to malfunction of the hardware of the system, instant action will be taken most of the time entailing the deactivation of the wearable device.

From a technical point of view, this level can be perceived as the one where the processor of the wearable system design detects a hardware malfunction. This detection can be a direct one when a part of the wearable device informs about its malfunction or an indirect one where the main processor infers the malfunction using its input data. In the first case, the hardware part has detected its malfunction and raises an interrupt to the main processor. The latter case refers to the scenario that the data gathered from a sensor is erroneous or out of normal range, for example, having a pH sensor reading value greater than 14, given a normal range of operation of 5.5–13. Finally, this is also the case when the sensors involved in the examination of the integrity of the system have normal values in terms of operational range but their values do not match, indicating that the system is not working under its reference value. The wearable device is turned off only in case that the detected malfunction is critical for the operation of the device especially in terms of whether the user is affected by the aforementioned malfunction. For example, supposing that the malfunctioning sensor input is correlated to the control of an actuator of the device, then the device must be turned off or operate in a safe mode to maximize its ability to protect the user.

7.3.2 LEVEL 2: WARNING GENERATION FOR EXTREME SENSOR VALUES

This is the first level where data are examined from a medical perspective. In this level, hardware malfunction is not considered, since it has been already checked in the previous level. Thus, entering level 2, we ensure that both the actuators of the medical device, if any, and its sensors operate appropriately. However, the data that are read from the sensors may be outside their normal operating range as it has been defined by medical experts. Appropriately, a warning is generated for the patient to contact his or her physician and the clinical back-end server is informed about the incidence. Given that the produced warning is of medical and not functional relevance, the device remains operational in order to continue monitoring the values of all sensors. In continuation to the example used in the previous level using the pH sensor in this level, a warning would be raised, in case of a measurement out of the normal range is registered, for example, pH less than 5.5 and greater than 9. The example is clearer in the following pseudo code snippet:

```
If (ph_measurement < 5.5) OR (ph_measurement > 9) then {
      Produce_warning_via_UIF;
      Upload_event_to_BE_server;
}
```

7.3.3 LEVEL 3: WARNING GENERATION FROM MEDICAL RULE ENCODING

In this level, the sampled measurement of a sensor is within normal operating range and thus it remains to be examined if its value indicates that the status of the monitored process is deteriorating. This is also a range of values provided by the medical experts for a subset of the available sensors. The outcome of this level is also a warning urging the patient to contact his or her attending doctor and an upload of the event to the back-end server. As explained in the description of the second fusion level, the device remains operational. In continuation of the pH sensor example, assuming that medical experts have characterized a pH subrange of 6.5–7.3 as the one indicating a noncritical situation of the process under monitoring, this is the case where a warning is raised when the measured values are within range of 5.5–6.5 or 7.3–9. In pseudo code format, the example is as follows:

```
If (ph_m > = 5.5 AND ph_m < = 6.5) OR (ph_m > = 7.3 AND
ph_m < = 9) then

{
      Produce_warning_via_UIF;
      Upload_event_to_BE_server;
}
```

An important comment is that an inspection of the two examples in levels 2 and 3 could raise the question why could they not be merged in a single check to produce a warning in a pH measurement less than 6.5. This is because levels 2 and 3 have different medical semantics, which is mandatory to be examined distinctly for the medical expert and the patient to have an accurate view of the status of the wound.

7.3.4 LEVEL 4: WARNING GENERATION BASED ON MACHINE-LEARNING INFERENCE

In the aforementioned levels, data manipulation was straightforward in the sense that individual sensor readings and ranges of values are examined. This level of data fusion utilizes more advanced computational techniques, in order to discover and utilize correlations of input sensor data that are difficult to be defined and analyzed without the use of information technology. More specifically, this level of data fusion uses machine-learning techniques for enabling the assessment of complex scenarios. We focus our analysis mainly on classification algorithms.

When a phenomenon is investigated, most of the times there is a set of input observations and a set of possible output classes, where each vector of measurements belongs to. The problem of classification resides in the creation of a mathematical model to classify a new vector of observations in one of the available sets. An essential part of learning algorithms for classification is the training phase. This is also a key element differentiating them from the formerly described levels of fusion. In this phase, the algorithms adjust their internal structure to fit the data that they need to classify. This is why these classification algorithms are regarded as a subset of machine-learning operations. After the training phase is completed, a new classifier is created and it can "decide" in which class a new input measurement vector belongs to. The training of the classification algorithm has two important features that sometimes render its execution and consequently the use of the classification algorithm impossible:

- The training phase requires as input a quite large and descriptive set of measurement vectors with the necessity for each one of these vectors to be labeled with the class it belongs to.
- The training phase can be a computationally intensive task, which means that it would be a highly time-consuming task to execute on a wearable device with limited computational capabilities such as a wearable medical device. This issue is mitigated by the proposed cooperative fusion approach, in which training is performed on a powerful computational device such as the computer server infrastructure of a clinical site, and according to the results of this offline training, the classification algorithms are incorporated into the software of the wearable medical device. Consequently, the actual classification/prediction phase is performed online on the wearable device since it is computationally light enough to be executed on an embedded device.

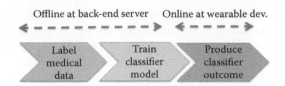

FIGURE 7.3   Cooperation between the offline labeling and training at the clinical back-end and the online prediction on the embedded wearable device.

Figure 7.3 depicts in an abstract manner the cooperative offline and online methodology. There are three major steps:

1. The medical data from the system sensors allocated to each data fusion engine are gathered in a central database. Afterward, the medical experts utilize their experience to label them, which means identifying what the status of the monitored process is in the case of each individual vector of sensor measurements.
2. After the labeling phase, the actual training of the candidate classification algorithms is performed. As mentioned, all the training activities take place in the back-end server. The efficiency of the trained model is cross-validated against characterized but unseen sets of real input vectors. This set is fed to the classifiers, and the error of their outcome is measured. In case of great prediction error, the training phase is executed once again using different parameters for the algorithm under training, until an acceptably small prediction error is achieved.
3. The trained classifier model proven to be superior in terms of predictive efficiency is forwarded/downloaded to the wearable device. The classification takes place during the actual operation of the wearable medical device. During its operation, in the event of a new sensor measurement, the newly acquired sensor data vector is classified using the preloaded algorithms and an assessment of the monitored process is acquired in real time.

7.4   DESCRIPTION AND CATEGORIZATION OF CLASSIFICATION ALGORITHMS

In a coarse manner, classification algorithms can be categorized in respect to the data set characteristics, for example, linear separable versus nonlinear separable data. Linear classifiers include classification algorithms that are able to manage data where all input vectors belong to classes that are distinct enough for a single line to separate them. Figure 7.4a illustrates such an example, while Figure 7.4b shows the case of a nonlinear separable data set [3]. The points in blue belong to class A while the points in red belong to class B. As depicted in Figure 7.4b, the distribution of data makes it infeasible to determine one line to separate the data in different classes. When working with greater dimensions, where we have input

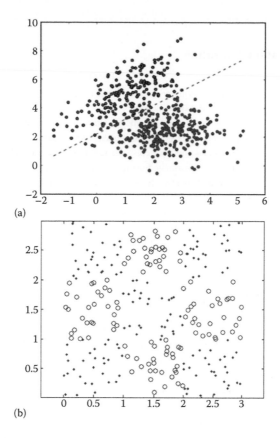

(a)

(b)

FIGURE 7.4 Linear versus nonlinear separable data set: (a) linear and (b) nonlinear. (From Theodoridis, S. et al. 2010. *Introduction to Pattern Recognition: A MATLAB Approach.* Academic Press.)

vectors with three or more data, instead of lines, hyperplanes are considered to distinguish between the different classes, but the classes are still perceived as linearly separable. It is important to state that a classifier can always make a classification decision. The goal is to discover and optimize the classifier that minimizes the classification error.

Table 7.1 summarizes the most well-known families of linear classification algorithms. In real-world applications, the data are multiparametric and nonlinear; thus, it is usually impossible to create a perfectly fitted linear model. In Sections 7.4.1 through 7.4.3, we examine in more detail nonlinear classifiers, taking into consideration both the requirements of the online classification algorithm in time and memory and its error at classifying the training data set as well as the testing data set. The three most effective algorithms for classification of nonlinearly separable data are described, namely, artificial neural networks, tree classifiers, and support vector machines (SVMs).

TABLE 7.1
Linear Classification Algorithms

Classifier	Description
Mahalanobis distance	All classes are equally probable. Data in all classes are described by Gaussian probability density functions. Mahalanobis distance from mean value of each class classifies the new point.
Naïve Bayes	The probability of a new measurement vector x to belong to class k is calculated. Vector x is classified to a class where it most probably belong to.
Least squares method	The classifier is estimated using the method of least squares estimation. A new measurement vector x is classified using the outcome of the solution of a linear system with weights w.
k nearest neighbors	Given a test vector x, the distance from its k nearest points is calculated, where k is a parameter of the classifier. Vector x belongs to the class that the majority of its k nearest points belong to.

7.4.1 ARTIFICIAL NEURAL NETWORKS

In artificial neural network classifiers [4], the core of the model is the neuron, which is a system with multiple inputs and one output. Each input is multiplied with a weight factor, computed during training, and the summation of these products with the addition of a bias number is directed to a mathematical function, called activation function, providing the output of the neuron. Common activation functions are the linear one, the hyperbolic tangent, and the logarithmic one. Figure 7.5a depicts the neuron as a mathematical construct [5].

The artificial neural network derives from the combination of individual neurons to a network topology in order for the final result to be produced. The input vector of data is directed to a number of neurons that belong to a so-called hidden layer. The output of these neurons is directed to the final layer where we have one or more neurons to produce the final result. When we have only one output neuron, we can perceive the result as the neural network trying to emulate a mathematical function with multiple inputs and one output. In the case of multiple-output neurons, the output of the ith output neuron can be perceived as the probability of the input to belong to class i. The training of the neural network refers to the definition of the weights and the bias of each neuron, both in the hidden and the final layer. There are several algorithms, supervised or nonsupervised, for the training of a neural network [4]. In the proposed data fusion framework, we utilized a supervised training strategy that fits well in training neural networks on the basis of prelabeled data.

In order to evaluate the algorithmic complexity of the online classification phase of an artificial neural network, we examine a network with N inputs, one hidden layer with H hidden neurons, and K output ones. This network configuration is the most common one, and the following analysis about complexity can be easily generalized for a network with more hidden layers or output neurons. In each hidden neuron, N multiplications and $(N + 1)$ additions are executed. For H neurons, we get $H \times N$ multiplications. In the output, there are H multiplications as the output of the hidden

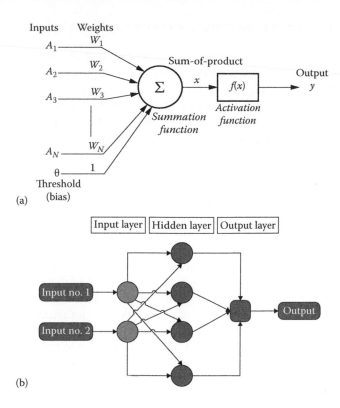

FIGURE 7.5 General structure of a neuron and a neural network: (a) neuron [5] and (b) neural network.

neurons is gathered and weighted in the output one. In total, we have $H \times N + H = H \times (N + 1)$ multiplications. Thus, in formal representation, the computational complexity of neural network with the aforementioned characteristics is $O(H \times (N + 1))$. There is no quantitative relationship between H and N in order to eliminate one of the two. In addition, it is pointed out that the former equation is the computational complexity for only one vector of measurements. In case of M vectors, the equation has to be multiplied by a factor of M.

7.4.2 Classification Trees

This nonlinear classification method utilizes a binary tree as a classifier [4]. Since this method is very common in many fields, the classification tree can also be found as a decision tree or a prediction tree. The core of the method is that each node of the tree makes a decision using a logical condition combining one or more of the input parameters. The tree is binary since the output of the logical operation can be either true or false. Each one of these results leads to another node or leaf of the tree. Using this rule, a path is followed inside the tree that leads to a specific leaf and that leaf defines the class the input vector belongs to. The training of the classification tree refers to the process of discovering the logical condition of each node.

As far as the computation complexity of the online decision tree prediction is concerned, it is difficult to define an exact equation owing to the difficulty of predicting a priori the height of the tree resulting from a given input data set. As a general rule, if the tree has a height of H value, then traversing through requires at most (worst-case complexity) H logical operations, which, in most practical cases, is not a computationally intensive task.

7.4.3 SUPPORT VECTOR MACHINES

In SVMs [4], the goal is to determine the optimum hyperplane to separate the points of different classes in a multidimensional space. The same applies for the two-dimensional space where the goal is to define the line separating different classes.

Figure 7.6 shows that the definition of the best separating line is not a trivial choice, as many lines may have the necessary qualifications to be chosen as the one [6]. At this point, the concept of the support vector is introduced. Support vectors are the data points closest to the optimum line separating the two sets. As a consequence, a change in the position of one support vector changes which line is considered the optimum. Support vectors are the critical elements of the training set. The goal of finding the optimum line is directly connected to finding the line with the maximum margin between the separating line and all the points of one class. In this way, the classes are structured with a maximized probability of a new point to be successfully classified.

Although a typical SVM classifier, as the one described, targets linearly separable data, it allows the incorporation of nonlinear kernels to be utilized in order to support classification of data that are not linearly separable. The main idea of utilizing a nonlinear kernel is to map the input data, via the means of a nonlinear function, to a space in which they can be linearly separable. In Figure 7.7, we show an illustrative example taken from Ref. [7] in which the input data belong to two different classes

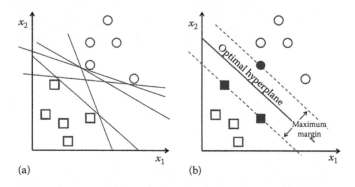

FIGURE 7.6   Example of (a) multiple separation lines and (b) the optimal hyperplane. (From Bradski, G. and Kaehler, A. 2008. Learning OpenCV: Computer vision with the OpenCV library. O'Reilly Media, Inc.)

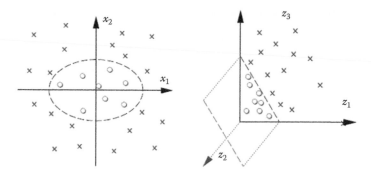

FIGURE 7.7   Application of a nonlinear kernel function to a nonlinear separable data set.

that can be separated with a cycle. If the data are transformed to polar coordinates, then a line is the optimum separator of the two classes and the SVM classification algorithm can be employed to discover that line. The transformation function is referred to as a kernel function and it is usually a nonlinear one. There are plenty of different choices for a kernel function such as a polynomial one with order greater than 1, *Gaussian radial basic function (RBF)*, and *hyperbolic tangent (tanh)*.

The choice of kernel function significantly affects the computational complexity of the SVM classifier. In the linear case, the computational complexity of calculating the class of one input vector is linear. In the nonlinear case, the input vector has to be transformed to another mathematical space, and this transformation imposes its own computational cost. Considering that all these calculations are floating point operations, the use of some kernel functions may affect performance. In any case, the computational complexity for one input vector is the same as the nonlinear case, multiplied with the complexity of the application of the kernel function to the input vector.

Figure 7.8a illustrates the outcome of the neural network classification algorithm with 10 hidden layers for the nonlinearly separable data set of Figure 7.7. Figure 7.8b illustrates the outcome of the decision trees classification algorithm for the nonlinearly separable data set of Figure 7.7. Figure 7.8c illustrates the outcome of the SVM classification algorithm for the nonlinearly separable data set of Figure 7.7.

7.5   CASE STUDY—DATA FUSION FOR WOUND MANAGEMENT

The data fusion scheme presented in Section 7.3 forms a generic framework that can be applied to a wound management and monitoring infrastructure that involves a number of sensors and the corresponding intelligence to provide a set of outputs either to inform a user or to control a number of actuators. Indeed, the aforementioned data fusion methodology has been applied within the SWAN-iCare* project [8] for wound monitoring and management infrastructure. Chronic wounds form an emerging hospitalization factor especially for elderly people. More than 10 million people in Europe

* SWAN-iCare. Available at http://www.swan-icare.eu.

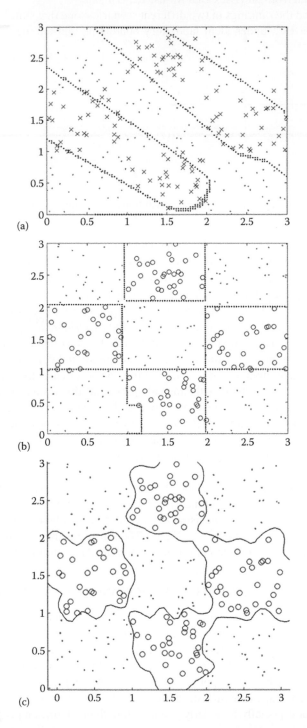

FIGURE 7.8 Classification of the test nonlinear separable data set using (a) neural networks, (b) decision trees, and (c) SVMs.

suffer from chronic wounds, a number of which is expected to grow because of the aging of the population. In order to address chronic wound management, a wearable device constantly monitoring the status of the wound and providing information on the healing process and early identification of wound deterioration can be proven extremely critical both for (i) improving the patient's quality of life, since the patient's need for hospitalization is minimized with a reassurance that the wound's condition is appropriately monitored, and (ii) minimizing healthcare costs and hospitalization, without sacrificing the quality of treatment. The ultimate goal of the project is to put together all the necessary components to develop a system of efficient ecosystem for chronic wound management on the basis of the medical concept of negative pressure wound therapy (NPWT) [9], in which negative pressure is applied on the wound to assist its healing process. In the core of the SWAN-iCare ecosystem, there is an embedded smart wearable device (SWD) to (i) monitor the biological parameters of the wound, (ii) combine them in order to assess the wound status, (iii) enforce and control the negative pressure therapy, and (iv) provide all these information to a back-end clinical server for further analysis by the healthcare experts.

Sections 7.5.1 and 7.6 analyze in more detail the instantiation schemas of the cooperative data fusion engines located at the patient's wearable device and at the clinical infrastructure.

7.5.1 Data Fusion Engine on WD

Table 7.2 summarizes the sensors directly connected to the SWD. There are sensors related to the pump (speed, current, and pressure), which is the only actuator of the SWD and the most critical element for enforcing NPWT. The rest of the sensors convey information about the status of the wound and are in close contact to it. The table lists the cases for which an alarm or warning should be generated from the data fusion engine of the SWD. It also summarizes the indicative range of values that could raise an alarm in the system. Table 7.2 reports for each sensor with medical relevance allocated to the SWD data fusion engine (i) the inferred wound assessment (i.e., negative pressure, inflammation, and infection), (ii) the normal operation range (column 3) to be used on data fusion level 2, (iii) the minimum and maximum values that trigger an alarm or warning generation (column 4) to be used on data fusion level 3, and (iv) the corresponding receiver per alarm (column 5).

Figure 7.9 shows how the original sensor fusion framework is instantiated as the data fusion engine of the WD. The pump sensors (i.e., pump pressure, pump motor speed, and pump motor current), the on-wound integrated sensors (i.e., MMPs, TNFα, pH, and wound temperature), and the SWD's activity monitoring sensor are accessed by the software running on the embedded wearable device and processed through levels 1–3 of the data fusion engine. The activity sensor and the on-wound integrated sensors are checked for malfunctions while the combined consideration of the reference value of the pump and the values of the pump sensors produce knowledge on both the state of the pump control and the pump-related sensors. Level 1 produces alarms, since it is directly related to the correct function of the device. These alarms are forwarded to the clinical back-end server via the wireless communication interface and to the user interface of the SWD, for the patient to be informed.

TABLE 7.2

Encoded Medical Conditions for Alarm and Warning Generation at the SWD Data Fusion Engine

List of Sensors	Frequency of Measurement	Normal Range or Range of Values	Send Alarm When below or above These Values		Alarm or Warning to Patient, Nurse, GP/Specialist
			Min	Max	
Negative Pressure on Wound					
Negative pressure	Continuously	25–125 mm Hg	25	125	Alarm to patient + nurse
Inflammation					
Wound temperature	Daily	20°C–25°C	<20	>25	Warning to nurse and specialist
MMPs 9	Weekly	1–20 µg/mL	No alarm	No alarm	No alarm
TNFα	Weekly	0–1500 pg/mL	No alarm	No alarm	No alarm
Infection					
pH	Daily	5.5–9	<6.5	>7.3?	Warning to nurse and specialist

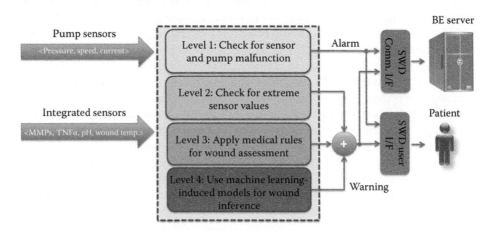

FIGURE 7.9 Data fusion engine at the SWD.

Levels 2 and 3 are used only for the measurements of the integrated sensors and produce conclusions that encode medical knowledge on the values of sensors, related to the state of deterioration of the patient's wound. The combined result of these two levels produces warning messages about the status of the patient's wound, and this warning is sent to the back-end server and UIF interface of the SWD following the same procedure as the alarms of level 1. Level 4 executes the classification algorithm of the device using the input of the integrated sensors. Its intended result is to assess the status of the wound. Provided that this assessment implies that the wound status is deteriorating, an appropriate warning (similar to the previous one) is forwarded to the back-end server.

7.6 DATA FUSION ENGINE ON THE BACK-END SERVER

The data fusion engine in the clinical back-end server aims at expanding the functionality performed to the SWD. This is achieved by aggregating the results and raw data coming from the device with data coming from a variety of other sensors apart from the on-wound ones. These sensors are related not only to the status of the wound but also to the entire clinical overview of the patient. More specifically, the available sensor devices fall into the following categories:

- *Stationary medical sensors*: They are available on the market and are able to fulfill specific medical needs. They include a *thermometer*, a *pulse oximeter*, and a *glucometer*.
- *External sensors*: This term refers to sensor devices that are developed to measure information related to the wound without being constantly attached to it. They are placed in contact to the wound only during the change of its dressing. Examples of these sensors are *transepidermal water loss devices*, *wound tissue impedance devices*, and *bacteria identification sensor devices*.
- *Wearable sensors*: Wearable devices refer to sensors that the patient carries with him or her, fixed either around a leg or insole. Examples include *dorsiflexion devices* and *insole pressure devices*.

Figure 7.10 presents the fusion levels of the corresponding data fusion engine in the clinical back-end:

Level 1: It concerns checking if there is a malfunction in one or more system components. In this, it is also taken into account how the components interact with each other and how the erroneous operation of the one affects

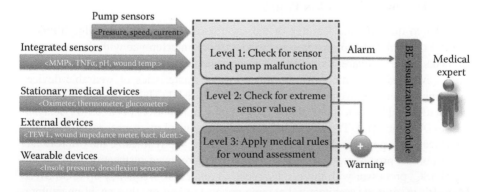

FIGURE 7.10 Data fusion engine at the clinical back-end server.

the others. The output of this level is an alarm shown to the medical expert through a specific user interface.

Level 2: It concerns checking of extreme sensor values (minimums and maximums).

Level 3: In this level, medical rules are checked concerning the progress and condition of the wound. The output of this level in combination with the output of Level 2 will generate a warning that will be presented in the user interface of the medical expert.

The approach followed in the clinical back-end is the same with the one described in Section 7.5.1. The size of data generated and stored in the back-end is considered important for the variety of results and alarms produced. Thus, the more data available, the more accurate and different results can be generated. The type of alarms or warnings depends also on the risk management analysis that identifies potential hazardous situations that may be caused by malfunctions or extreme sensor values. When an alarm or warning is generated, it is also foreseen that an SMS (short message service) or mail will be sent to the medical experts. This is configurable and can be customized based on medical experts' personal needs. Additionally, a daily report sent to the doctor through mail is generated, including all alarms/warning as well as trends for his or her assigned patients.

7.7 EXPERIMENTAL SETUP AND ANALYSIS

In this section, we experimentally analyze and evaluate the effectiveness of the proposed data fusion algorithms. Taking into account that the SWD is an embedded platform with limited computational resources, the ported software is expected to exhibit differing performance figures compared to a general-purpose machine. Given this fact, we developed an SWD emulation infrastructure running on an embedded platform with relatively limited processing power and memory size, in order to enable an early performance analysis.

7.7.1 HARDWARE EMULATION PLATFORM

In order to quantify the features of the sophisticated classification algorithms of Level 4 of the data fusion scheme and especially its interference with the most critical tasks of the software of the SWD, a co-designed hardware/software platform has been developed [10] emulating the functional characteristics of wearable devices targeting wound management and monitoring. Figure 7.11 shows the devised emulation platform. The basic element of the HW emulation platform is a Xilinx Spartan-III FPGA device [11]. The FPGA instantiates the control microprocessor and every interface of the peripheral devices is connected to it. We synthesize the MicroBlaze, a soft-core IP processor, provided by Xilinx. It is a reduced instruction set computing (RISC) processor with three-stage pipeline and clock frequency up to 50 MHz. MicroBlaze supports architectural parameters customization, thus enabling exploration of differing design configurations to be performed, in order to tailor the design

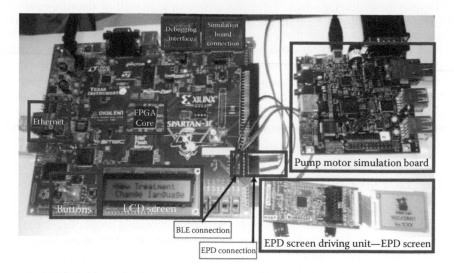

FIGURE 7.11   FPGA emulator hardware platform.

to the characteristics of the application's SW components. It forms a quite good match considering the requirements of a wearable medical device since it is more powerful than typical microcontrollers and has similar architectural features to ARM processors [12] that are frequently used in the design of portable devices. The on-board Ethernet module of Spartan-III has been used to implement the communication of the embedded device with clinical back-end server. The LWIp TCP/IP protocol stack [13] was used and the processing of the packets was a task that was handled by the main processor. The server was implemented on a desktop computer where the medical device uploads its information.

A Bluetooth Low Energy module has been allocated for communicating wound sensor data with the main processor. Wireless interface has been selected instead of a wired one, since lack of wires enables the design of a more comfortable device. In addition, use of wireless communication technology provides the ability to create different configurations of used sensors according to the individual case of each user. Table 7.2 reports the in-wound integrated sensors that are developed within Swan-iCare to monitor the wound status.

Regarding the user I/F, an ePaper Display (EPD) was connected through a universal asynchronous receiver/transmitter (UART) interface. Specifically, we integrated the hardware and firmware components of the AdapTag development kit [14] from Pervasive Displays. The EPD shows information about the status of the wearable device and messages about the evolution of the treatment. We used the 1.44″ EPD panel that satisfies the need for easy-to-read messages.

In the proposed HW platform, the real-time control of the pump is performed through the speed regulation of the motor driving the pump. A model of the pump in the time domain was created in MATLAB® [15] using nominal values from an actual DC motor and then a proportional–integral (PI) controller was configured to function as the controller of this motor. PI control was used in order to ensure the stability of the control by sacrificing settling time. This choice was made on the premise that it is important to ensure that the pump will not momentarily assert great pressure on the wound, which could probably damage the tissue on and around the wound. To achieve real-time simulation of both the motor behavior and the controller response, the model of the motor was implemented on a Beagle development board [16] external to the FPGA. From a co-design point of view, this offers great flexibility since the external development board can be used to implement any model of any device under control with relative low effort and it is up to the software designer to satisfy the constraints of the control algorithm in the software managing the wearable medical device.

7.7.2 Data Set Generation

In order to enable an early evaluation of the various data fusion schemes, an artificial data set has been generated. According to the normal operating range of each sensor, we randomly sampled the vector space defined by sensor value ranges, assuming that the points abide to the Gaussian statistical distribution. After a complete set of measurements was created, it was labeled according to whether there is value inside the range of values that are medically important and a warning is created for the attending doctor, according to Table 7.2. This input data set consists of 200 vectors of measurement of all integrated sensors excluding the bacteria identification sensor since it is more of a binary sensor indicating either the existence or the absence of a population of some specific bacteria.

The input data set of the 200 vectors was used to train the nonlinear classification algorithms. This train took place in a desktop personal computer using MATLAB R2013 [15]. When a new emulation instance of the SWD is initiated, during the initialization of the program, according to the data fusion scheme that is being tested, these files are read to replicate the structure of the classifier inside the emulation software. Finally, using the same distribution for the operating range of the integrated sensors, a set of input vectors was created to act as a test data set during the emulation of the SWD. This data set consists of 50 input vectors. The size of both data sets was arbitrarily chosen and the testing infrastructure is designed in such a way that it is easy and fast to create input data sets of any number of vectors if this is necessary in future exploration of the functionality of the device. Whenever the SWD emulator asks for a measurement from a specific sensor, the corresponding function of the emulation software is executed and it returns the appropriate value read from the test data set input file.

We focus our analysis on the machine-learning algorithms belonging in level 4 of the data fusion framework. This is because while the implementation decisions are more or less straightforward in levels 1–3 of the data fusion scheme, in level 4, the developer has to face the problem of selecting the best candidate in terms of classification efficiency and computational burden from a set of available data fusion algorithms.

7.7.3 Quality Analysis of Data Fusion Engine

In this section, we analyze the efficiency of the employed classification algorithms in terms of classification/prediction error.

Case 1: Linear separable data sets: We considered that the input data set consists of two classes with 350 vectors each one, which leads to a 700-vector data set. The vectors are five-dimensional resembling the number of employed sensors on the SWD (Table 7.2) derived from a Gaussian distribution. The data set used for testing is 300 vectors created from the same distributions as the training data and equally distributed to the two classes. The following classifiers were used:

1. Mahalanobis distance classifier
2. Naïve Bayes classifier
3. Perceptron neural network (only one neuron)
4. Least squares classifier
5. SVM linear classifier
6. Prediction trees
7. k nearest neighbors classifier

Figure 7.12 illustrates the classification error of each classifier in the classification of the 300-vector data set. The graph shows that the sophistication of a classifier is a first step toward less classification error, since the SVM classifier is the classifier with the least error. However, all classifiers exhibit a relatively big error, more than 15% in all cases, which would be unacceptable in a real medical application. Would that be the case, a greater fine-tuning of the classifiers should have been employed to better fit the input data set or more complex fusion algorithms should have been used.

Case 2: Nonlinear separable data sets: In this case, the input data set consists of 280 two-dimensional input vectors while the data set to be classified consists of 180 vectors. The tests were conducted as described in the previous section. Figure 7.13 shows the classification error on the test data set among the three nonlinear classifiers. In this case, the decision tree seems to significantly prevail over the other two algorithms.

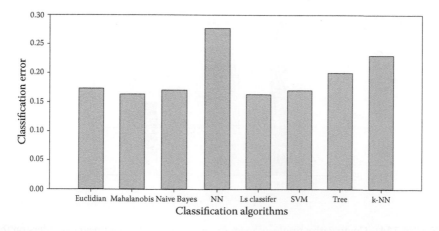

FIGURE 7.12 Comparative classification error for a linear separable case.

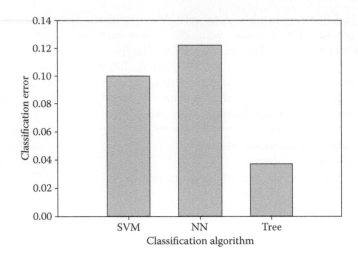

FIGURE 7.13 Comparative classification error for a nonlinear separable case.

All measurements have been acquired during the execution of the emulator of the SWD. The measurements of execution time were conducted using high-resolution timers that return the time of system with granularity of microseconds. Only the operation of the classification algorithm is measured; no other part of the emulation software is measured including the central timer of the program, which is temporarily moved to lower priority during the execution of the algorithm so that it does not interfere with the measurement. Figure 7.14 depicts the measured values in microseconds.

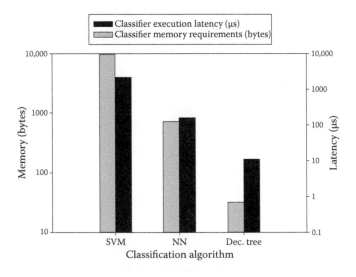

FIGURE 7.14 Comparative results of memory requirements for nonlinear classifiers on the SWD emulation platform.

The presented values are the mean values of 250 executions of each algorithm during five emulation sessions. The observed results coincide with one's expectations considering the structure of the classifiers. More specifically,

1. SVM is the classifier that executes the most calculations since it computes a nonlinear transformation and then it solves a relatively big (in terms of coefficient number) linear system.
2. Neural network comes second in complexity since it executes a small number of linear calculations. This fact explains the big gap in execution time compared to the SVM classifier.
3. The decision tree executes only comparisons that are simple logical operations. This justifies its very small execution time. In fact, many times, its execution is less than 1 μs, resulting in zero measured time difference by the emulator. (This is technically feasible since the embedded board functions at a frequency of 1 GHz.)

Figure 7.14 illustrates the calculated memory requirements of the classification algorithms. The required memory of each classification is the one statically allocated and can be measured by inspection of the code.

In direct analogy to execution time measurements, the complexity of the investigated algorithm severely affects its memory requirements. Thus, as expected, the SVM classifier is the one with the highest need for memory, owing to storage of the coefficients used in the linear system that it computes. The neural network classifier has significantly less requirements compared to SVM and significantly more requirements compared to the decision tree. The network only needs to store the weights and the bias values of its neurons while the tree only needs to store the values in its nodes, which are used for comparisons. To readers experienced in using trees, this may seem odd, since in most programs every node holds information about the following nodes as well. However, in our case, the structure of the tree is hardcoded to eliminate its need for a bigger memory size and dynamic allocation of data. Conclusively, the memory of the presented classifiers even in the worst case is less than 10 kB, a result acceptable to the vast majority of modern embedded platforms.

The emulated SWD was utilized to analyze the impact of differing architectural decisions on the timing of the different tasks and the SW ability to meet the necessary time constraints of the medical devices. Architectural decisions were explored regarding (i) the memory architecture, that is, instruction and data cache configuration, and (ii) the inclusion/exclusion of the a floating point unit (FPU), across embedded application instances with differing machine-learning algorithms for the data fusion engine, that is, neural networks, SVMs, and decision trees. Figure 7.15 depicts the impact of cache size (instruction and data) and FPU allocation on the performance of the fusion engine. As shown, decision trees form the most efficient decision regarding performance. The existence of an FPU in the microprocessor reduces the execution time of the algorithms operating on floating point data, such as SVMs and neural networks. Decision trees are not affected since their code structure is based on branch instructions, which do not require complex FP operations to benefit from the FPU. In contrast to the FPU, the cache memory size should be carefully chosen

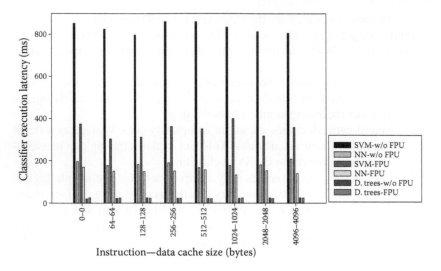

FIGURE 7.15   Architectural configurations' impact on performance.

in order to speed up the execution, since the data access patterns in memory can be such that the average execution is increased even compared to the design with no cache memory, for example, 256 cache size configuration for the SVM without FPU.

7.8   CONCLUSIONS

In this chapter, we presented cooperative data fusion as an effective strategy to manage the analysis, correlation, and inference requirements found in modern multi-source wearable sensor infrastructures. The allocation of multiple cooperative fusion engines in computational nodes with differing performance capabilities enables a more effective and scalable deployment than a completely centralized one. In addition, by abstracting the semantic functionality of the data fusion engine, we presented a multi-level generalized data fusion scheme that can be customized accordingly to different use cases and input streams. We show that such a data fusion scheme is realistic and already adopted for the data aggregation and reactive analysis of a remote wound monitoring and assessment healthcare system. Extensive experimental analysis for the emulated wound management wearable device analyzes the trade-offs in QoS (quality of service) levels, resource requirements, and performance figures, thus enabling the prospective designer/programmer to guide through the differing design decisions.

REFERENCES

1. Physionet. 2012. MIT-BIH Physionet Database, Cape Town, South Africa. Available at http://www.physionet.org/physiobank/database.
2. Shoaib, M., Jha, N.K., and Verma, N. Algorithm-driven architectural design space exploration of domain-specific medical-sensor processors. *Very Large Scale Integration (VLSI) Systems, IEEE Transactions on* 21.10 (2013): 1849–1862.
3. Theodoridis, S., Pikrakis, A., Koutroumbas, K., and Cavouras, D. 2010. *Introduction to Pattern Recognition: A MATLAB® Approach*. Academic Press.

4. Abu-Mostafa, Y.S., Magdon-Ismail, M., and Lin, H.-T. 2012. *Learning from Data.* AMLBook.
5. Kiyoshi, K. Artificial Neuron with Continuous Characteristics. 17 June 2000 Web.
6. Bradski, G. and Kaehler, A. 2008. Learning OpenCV: Computer vision with the OpenCV library. O'Reilly Media, Inc.
7. Thornton, C. 2011. Machine Learning-Course website. Available at https://scholar.google.gr/scholar?q=Chris+Thornton+svm&btnG=&hl=en&as_sdt=1%2c5.
8. SWAN-iCare. Available at http://www.swan-icare.eu.
9. Texier, I., Xydis, S., Soudris, D., Marcoux, P., Pham, P., Muller, M., Correvon, M. et al. 2014. SWAN-iCare project: Toward smart wearable and autonomous negative pressure device for wound monitoring and therapy. In *Proceedings of IEEE International Workshop on Smart Wearable and Autonomous Devices for Wound Monitoring*, 2014 EAI 4th International Conference, pp. 357–360.
10. Tsoutsouras, V., Xydis, S., and Soudris, D. 2014. A HW/SW framework emulating wearable devices for remote wound monitoring and management. In *Proceedings of IEEE International Workshop on Smart Wearable and Autonomous Devices for Wound Monitoring.* 2014 EAI 4th International Conference, pp. 369–372.
11. Xilinx, Inc. Available at http://www.xilinx.com/products/silicon-devices/fpga/spartan-3.html.
12. Advanced Risc Machines (ARM), Inc. Available at http://www.arm.com.
13. lwIP—A Lightweight TCP/IP stack. Available at http://www.xilinx.com/ise/embedded/edk91i_docs/lwip_v2_00_a.pdf.
14. Pervasive Displays, Inc., Available at http://www.pervasivedisplays.com/kits/adapTag.
15. MathWorks MATLAB. Available at http://www.mathworks.com/products/matlab/.
16. Beagle Board. Available at http://beagleboard.org/beagleboard-xm.

8 Energy-Efficient High Data Rate Transmitter for Biomedical Applications

Chun-Huat Heng and Yuan Gao

CONTENTS

8.1 INTRODUCTION

Recent development on wireless medical applications, such as wireless endoscopy and multichannel neural recording integrated circuit (IC), has spurred the need for energy-efficient high–data rate transmitter. For example, wireless endoscopy with an image resolution of 640 × 480, 6 fps, and 8-bit color depth requires a few tens of megabits per second if on-chip image compression is not available. Similar requirement applies to a 256-channel neural recording IC with 10-bit resolution and a sampling rate of 10 kS/s. Such applications are often characterized by asymmetric data link as shown in Figure 8.1 where high–data rate uplink is required to upload critical biomedical data and low–data rate downlink is used only for configuring the implanted device. Therefore, complex modulation, which supports high data rate, does not necessarily lead to power-hungry solution if power-efficient transmitter architecture can be realized as the complex receiver function is usually implemented off-body without stringent power constraint. In addition, it has been found

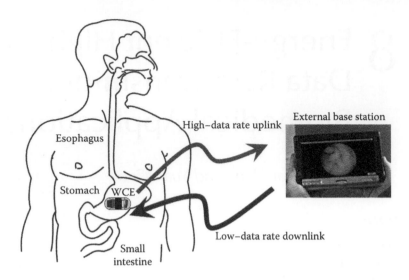

FIGURE 8.1   Asymmetric data link.

that transmission frequency below 1 GHz is generally preferred for such applications owing to smaller body loss (Kim et al. 2010). As this frequency range is often crowded, complex modulation with better spectrum efficiency would thus be more favorable.

The conventional mixer-based or phase-locked loop (PLL)–based approach, which supports phase shift keying (PSK) or quadrature amplitude modulation (QAM), does not lead to low-power solution. They are often limited by the need of power-hungry high-frequency building blocks, such as mixer, radio-frequency (RF) combiner, high-frequency divider, and PLL. In this chapter, we will discuss the choice of modulation and examine the various proposed techniques that can eventually lead to an energy and spectral efficient high–data rate transmitter.

8.2   MODULATION

Figure 8.2 shows the evolving trends on adopted modulations for transmitters at the industrial, scientific, and medical (ISM) band, millimeter-wave band, and sub–1-GHz band. Because of the improved technology and the demand on higher data rate over limited spectrum resource, designs are moving from simple modulation such as on–off keying (OOK) and frequency shift keying (FSK) toward more complex and spectral efficient modulation such as PSK or QAM. As shown in Figure 8.3, by moving from FSK to 16-QAM, the spectral efficiency can be improved by four times. The relatively high side lobe can be suppressed if pulse shaping is adopted as shown in Figure 8.4 to improve the spectral efficiency further. Nevertheless, there are a few requirements for implementing complex modulation and pulse shaping. First, accurate phase is needed, which demands accurate frequency and phase generation. This is conventionally accomplished with PLL. Second, pulse shaping can be achieved with either mixer-based approach or polar transmitter, which generally

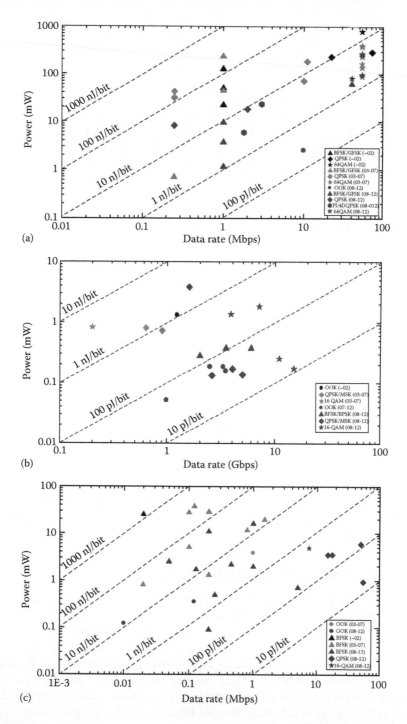

FIGURE 8.2   Transmitter evolution trend for (a) ISM band, (b) millimeter-wave band, and (c) sub–1-GHz band.

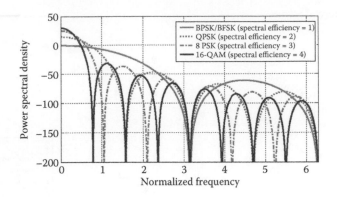

FIGURE 8.3   Spectral efficiency for different modulation.

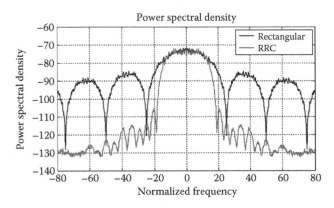

FIGURE 8.4   QPSK modulation with and without band shaping.

leads to complex architecture and higher power consumption as mentioned earlier. In Section 8.3, we will examine a few proposed techniques that can lead to simplified architecture and achieve low-power consumption.

8.3   FREQUENCY AND PHASE GENERATION

PLL is widely adopted for frequency and phase generation. However, because of the need of a high-frequency divider, charge pump, and so on, it is not a power-efficient solution. In addition, for the LC oscillator, multiphase is obtained by operating at twice larger the desired output frequency and dividing down the output through a high-frequency divider, which incurs more power penalty because of the higher operation frequency. On the other hand, the ring oscillator (RO) can readily provide the desired output phase but suffer from poorer phase noise. Here, we proposed two energy-efficient ways of achieving accurate frequency and phase generation without the aforementioned issues.

8.3.1   LC-BASED INJECTION-LOCKED OSCILLATOR

The injection-locked oscillator (ILO) has been found to be an energy-efficient way of obtaining accurate frequency reference without the need of additional components, such as frequency divider, phase frequency detector, and charge pump. Hence, it can lead to an energy-efficient solution for frequency carrier generation. Fundamentally, it behaves like a first-order PLL where the free-running frequency of the LC oscillator will lock to the nth harmonic of the injected reference frequency. Here, we proposed an efficient way of obtaining the desired multiphase output without resorting to higher operating frequency and divide down method. As shown in Figure 8.5, depending on the free-running frequency of the LC tank, the output phase exhibits a certain relationship with respect to the nth harmonic of the injected reference (Razavi 2004). For example, if the free-running frequency of the LC tank is higher than the nth harmonic of the injected reference, the output phase will be lagging the injected reference. On the other hand, if the free-running frequency is lower, the output phase will be leading the injected reference. This observation allows us to manipulate the free-running frequency of the LC oscillator to obtain the desired output phase. As shown, an output phase with ±45° can be obtained through this manner, and the remaining ±135° can be obtained by swapping the output phase to introduce the 180° phase shift (Diao et al. 2012). Finer output phase resolution can be achieved by using a switched capacitor bank to obtain fine frequency tuning for the LC oscillator.

From the aforementioned description, it is obvious that the frequency tuning will directly affect the phase accuracy and amplitude variation, which, in turn, will determine the resulting error vector magnitude (EVM) as follows:

$$\mathrm{EVM}(\%) = \sqrt{\left(\frac{\Delta M}{OI}\right)^2 + (\sin \phi)^2} \times 100\% \approx \sqrt{\left(\frac{\Delta M}{OI}\right)^2 + \phi^2} \times 100\%, \qquad (8.1)$$

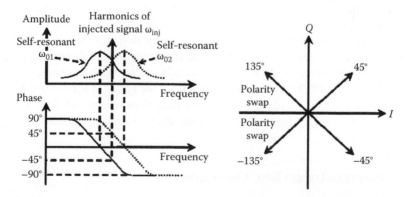

FIGURE 8.5   Phase modulation through modifying self-resonant frequency of LC tank.

where ΔM and ϕ are the resulting magnitude and phase deviations. It has been shown in Diao et al. (2012) that the oscillator output amplitude (V_{osc}) and phase are related as follows:

$$|V_{osc}| = \frac{I_{inj} \times \omega_{inj} L \sqrt{1 + \phi^2}}{\sqrt{\left(1 - \frac{\omega_{inj}^2}{\omega_0^2}\right)^2 + (\omega_{inj} L)^2 \left(\frac{1}{R_p} - g_m\right)^2}}, \tag{8.2}$$

$$\frac{\omega_0 - \omega_{inj}}{\omega_0} = \frac{1}{2Q} \cdot \frac{K \cdot \sin(\pm 45° + \phi)}{1 + K \cdot \cos(\pm 45° + \phi)}, \tag{8.3}$$

where I_{inj} is the injection current amplitude, ω_{inj} is the injected nth harmonic frequency, ω_0 is the free-running frequency of the oscillator, L is the inductance of the LC tank, R_p is the equivalent LC tank parallel resistance, g_m is the transconductance of the cross-coupled pair transistor of the LC oscillator, Q is the quality factor of the LC tank, and K is the injection strength given as

$$K = \frac{I_{inj,Nth}}{I_{osc}} = \frac{4}{N\pi} \frac{I_{inj}}{I_{osc}} = \frac{4\alpha}{N\pi}, \tag{8.4}$$

where I_{osc} is the oscillator current.

On the basis of Equations 8.1 through 8.4, the resulting EVM for a given phase error can then be obtained and is plotted in Figure 8.6. As shown, for EVM to be smaller than 23%, a phase error as large as 10° can be tolerated. Also, from the studies, the impact of magnitude error on EVM is much smaller compared to phase error as illustrated in Figure 8.7.

To determine the design requirement on the capacitor bank, the phase versus frequency characteristic can be obtained based on Equation 8.3 and is shown in Figure 8.8. As illustrated, a larger K allows the desired phase to spread over a larger frequency range. This will relax the capacitor bank design and is thus desirable. Figure 8.9 shows the design choice on K and the needed smallest unit capacitance per degree phase resolution. To avoid false locking to wrong harmonics, a locking range smaller than 100 MHz is desired. On the other hand, to achieve higher data rate (tens of megahertz), a locking time smaller than 10 ns is required. This limits the range of K to 0.55–0.72, which translates to the smallest unit capacitance per degree phase of 3–5 fF/°.

8.3.2 INJECTION-LOCKED RING OSCILLATOR

For the sub–1-GHz regime, RO is a good candidate that offers readily available multiphase and compact area. However, it generally suffers from poor phase noise and

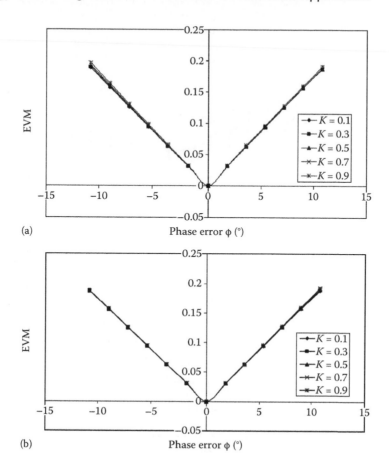

FIGURE 8.6   EVM versus phase error (ɸ) at (a) +45° and (b) −45°.

thus poorer EVM performance. Subharmonically, the injection-locked ring oscillator (ILRO) offers a compact and energy-efficient way of generating accurate output
frequency with multiphase output and good phase noise. As shown in Figure 8.10,
the main reason for the poor phase noise performance of RO is the jitter accumulation along the delay chain. By subharmonically injecting a clean reference signal, the
delay jitter is cleaned up during every reference period and it eliminates jitter accumulation. This leads to good output phase noise, which generally follows the phase
noise of the clean input reference even when a noisy RO is employed. Nevertheless,
there are a number of factors that will affect the resulting phase accuracy and thus
the EVM performance, which need to be carefully examined.

As shown in Figure 8.10, the output phase edge is only cleaned up at the beginning of each reference period. For the remaining interval, the oscillator will oscillate
at its own free-running frequency. Hence, a small phase error appears at the end
of each oscillator output period. This is equivalent to having a phase modulation
where the phase error accumulates over the entire interval and gets corrected at the

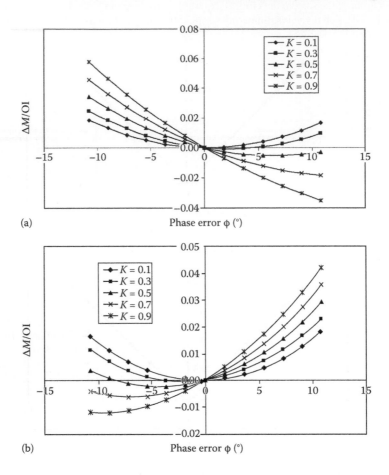

(a)

(b)

FIGURE 8.7   Relative magnitude error versus phase error (ϕ) at (a) +45° and (b) −45°.

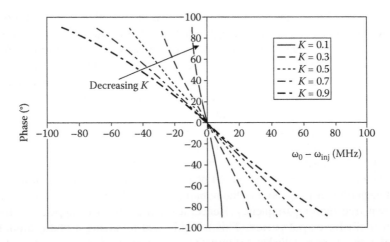

FIGURE 8.8   Phase characteristics for $\omega_0 - \omega_{inj}$ for IL-VCO.

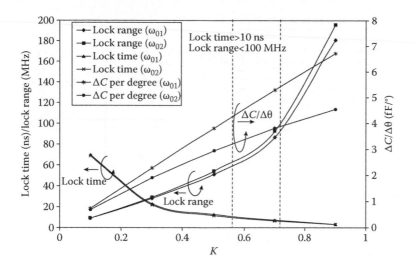

FIGURE 8.9   Choice of K.

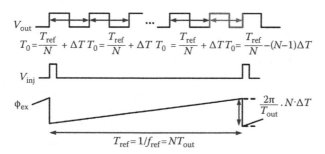

FIGURE 8.10   Systematic phase error accumulation.

beginning of the next reference period (Izad and Heng 2012b). By modeling this systematic phase error as phase modulation using a sawtooth waveform, we obtain

$$\theta_{\text{error,rms}} = \sqrt{\frac{1}{T_{\text{ref}}} \int_{T_{\text{ref}}} \left(\phi_{\text{ex}}(t) \right)^2 dt} = \frac{2\pi}{T_{\text{out}}} \cdot \frac{N \cdot \Delta T}{\sqrt{3}} = \frac{2\pi \cdot N \cdot \Delta f}{\sqrt{3} f_0}, \qquad (8.5)$$

where f_0 is the free-running frequency of the RO and Δf is the frequency deviation between the nth harmonic of injected reference and f_0. Using Equation 8.5, the desired tunable frequency resolution can be found for a given EVM. In fact, it is also found (Izad and Heng 2012a) that the reference spur level at the output is directly related to the resulting frequency deviation as follows:

$$\text{Spur (dBc)} = 20\log\left(\frac{N\Delta T}{T_{\text{out}}} \right) = 20\log\left(N \frac{|f_{\text{out}} - f_0|}{f_0} \right) = 20\log\left(N \frac{|\Delta f|}{f_0} \right) \qquad (8.6)$$

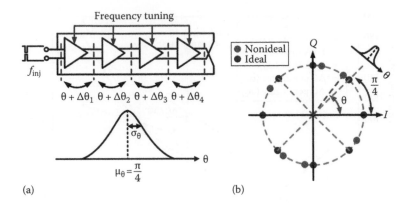

FIGURE 8.11 (a) Random mismatch between delay cells in the oscillator. (b) Effect of random phase mismatch on the constellation.

Hence, the measurement of the resulting reference spur level can serve as an indicator for the achieved frequency deviation and thus the systematic phase error.

The other factor that affects the EVM performance is classified as random phase error, which is mainly caused by device noise and mismatch. Because of the injection-locking mechanism, the RO output phase noise will follow the injected reference closely. Within the locking range, the output phase noise can be modeled as $L_{inj} + 20\log(N)$, where L_{inj} is the phase noise of the injected reference. The total root-mean-square (RMS) phase noise can be found by taking the root of the total integrated phase noise across the entire frequency band. To minimize its impact, low noise–injected reference and smaller N should be adopted in the design.

Because of mismatch between delay cells, the delay introduced by one stage can differ from another stage. For four-stage RO, each stage provides a mean phase step of $\pi/4$ (delay of $T/8$) with standard deviation of σ_θ. This mismatch manifests itself as a distortion of the constellation. In Section 8.4, we will look at ways to minimize this mismatch. The effect of these phase errors on constellation is shown in Figure 8.11.

8.4 BAND SHAPING AND SIDE-BAND SUPPRESSION

The QAM modulation can be described with the following equation:

$$s(t) = A(t)\cos(\omega t + \varphi(t)) = A_I(t)\cos(\omega t) + A_Q(t)\sin(\omega t) \tag{8.7}$$

As illustrated in Equation 8.7, the middle expression is the basis for polar architecture where fractional-N PLL is adopted to provide the carrier with arbitrary phase output and the supply-modulated power amplifier (PA) is adopted for amplitude modulation. On the other hand, the right expression is the basis for a mixer-based approach where the in-phase and quadrature-phase components [$A_I(t)$ and $A_Q(t)$] are generated at baseband, upconverted through a mixer, and summed through an RF combiner and PA. To avoid the issues faced by these two architectures, we adopt direct quadrature modulation at PA, which can be considered as a hybrid approach of the former two

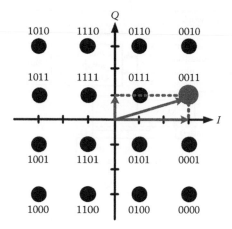

FIGURE 8.12 Constellation of 16-QAM.

architectures. First, the in-phase and quadrature-phase RF components [$A_I(t)$cosωt and $A_Q(t)$sinωt] are generated by driving PA with variable current strength from the oscillator multiphase output. The variable current strength is obtained by activating different numbers of transistors that get connected to a specific output phase. This eliminates the need of complex phase generation architecture and upconversion mixers. The summing of two components is then performed at the current domain, which can accommodate high-frequency operation. The idea is best illustrated in Figure 8.12. For example, to achieve the constellation point corresponding to (0011), we can combine φ_0 with 3× amplitude and φ_{90} with 1× amplitude. The concept can be easily extended to enable band shaping by providing multiple amplitude levels for the four-phase outputs. This will enable the fine phase and amplitude tuning required for band shaping as illustrated. From system simulation, it is determined that 5-bit amplitude control per phase is needed to achieve more than 38 dB side lobe suppression.

8.5 ENERGY-EFFICIENT TRANSMITTERS

In this section, we examine three energy-efficient architectures that have incorporated the proposed techniques mentioned earlier. The design insights associated with the circuit implementation will also be discussed.

8.5.1 QPSK AND O-QPSK TRANSMITTER BASED ON ILO

The energy-efficient quadrature phase shift keying (QPSK)/offset quadrature phase shift keying (O-QPSK) transmitter based on ILO is shown in Figure 8.13. The proposed architecture consists of only an ILO, a polarity swap circuit, a buffer, and a mapping circuitry, which transforms the input I and Q signals to the corresponding output phases. By eliminating the multiphase PLL and operating the VCO at the desired output frequency, the power consumption can thus be further reduced. As explained earlier, by controlling the free-running frequency of the LC tank to be lower than the harmonic frequency of the injected signal (solid lines), the locked LC

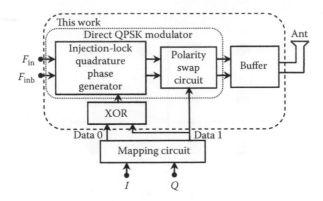

FIGURE 8.13 Block diagram of proposed QPSK/O-QPSK transmitter.

tank signal leads the harmonic of the injected signal by 45°. On the other hand, by making the free-running frequency higher than the harmonic of the injected signal (dotted lines), the locked LC tank signal lags behind the harmonic of the injected signal by 45°. Therefore, by changing the free-running frequency of the LC tank, we can create a phase difference of 90° in the output signal. The self-resonant frequency of the LC tank can be easily modified through capacitor bank switching. To generate all the four phases required for QPSK/O-QPSK modulation, an additional polarity swap circuit is employed to introduce 180° phase shift to the output signal. By employing both the capacitor bank and the polarity swap circuit, we could thus realize +45°, −45°, −135°, and +135° phase shifts required for QPSK/O-QPSK modulation.

The detailed circuit is shown in Figure 8.14. The ILO consists of a symmetrical NMOS cross-coupled pair (NM$_2$, NM$_3$), an LC tank incorporating a center-tapped differential

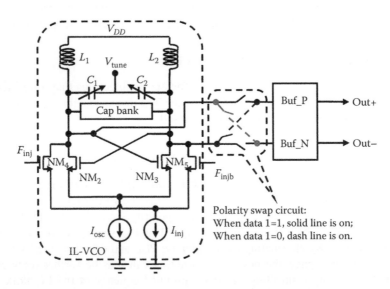

FIGURE 8.14 Schematic of the proposed QPSK/O-QPSK transmitter.

inductor ($L_1 + L_2$) and a capacitor bank, a differential pair transistor (NM$_4$, NM$_5$) for signal injection, and tail currents (I_{inj} and I_{osc}). F_{inj} and F_{injb} are the differential-injected signals. The free-running LC VCO has a free-running frequency (ω_0) centered around the targeted carrier frequency (ω_c). The designated harmonic of the injected signal (ω_{inj}) is chosen to be the same as ω_c. The free-running frequency can be changed by switching the capacitor bank to generate the desired phase shift as explained earlier. In this design, I_{inj} and I_{osc} are chosen on the basis of the desired K value mentioned earlier. The differential inductance is chosen to maximize its quality factor and parallel tank resistance (R_p) at around 915 MHz. A large R_p will reduce the tank current required for larger voltage swing and lead to lower power consumption. On the basis of momentum simulation, a differential inductor with a value of 20.8 nH and a Q factor of 5 is designed.

The schematic of a capacitor bank is shown in Figure 8.15a. To cover full ±90° phase range with the chosen 10° phase resolution, a 6-bit binary capacitor bank is required. The resonant frequency can be tuned to ω_{01} and ω_{02} by setting the control words B$_{-\omega 01}$[5:0] and B$_{-\omega 02}$[5:0], respectively. To avoid loading the tank, the Q factor of each switch-capacitor unit within the capacitor bank is designed to be around 45. This ensures that the total tank Q is limited by inductor.

The switch-capacitor unit is shown in Figure 8.15b. It consists of MIM capacitors C_1/C_2, shunt switches NM$_1$/NM$_2$, and a series switch NM$_3$, all controlled by B_N. NM$_1$/NM$_2$ will set the DC bias to ground when activated and can be kept small. The Q is mainly determined by series switch NM$_3$. This configuration halves the series resistance and allows smaller NM$_3$ to be used. This leads to smaller parasitic

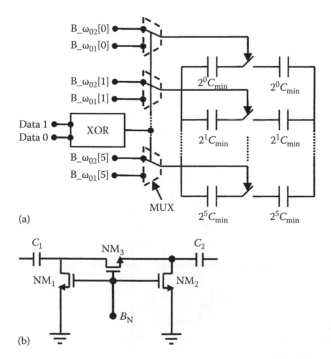

(a)

(b)

FIGURE 8.15   (a) Switched capacitor bank and (b) switched capacitor unit.

capacitance and larger C_{max}-to-C_{min} ratio. It has been shown in Diao et al. (2012) that getting absolute output phases of ±45° is not necessary in ensuring QPSK/O-QPSK with good EVM. This is because the resulting magnitude errors attributed to the phase deviations from ±45° are usually less than 2% and do not affect the EVM significantly. Hence, free-running frequencies corresponding to the output phase of ~+60° and ~−30° can be chosen as long as they exhibit a phase difference close to 90°. This helps relax the capacitor array design as larger unit capacitance is allowed, which results in larger frequency step size in self-resonant frequency tuning.

Gray coding is adopted for the phase modulation. Phases of 45°, −45°, 135°, and −135° correspond to Data0 and Data1 of "00," "10," "01," and "11," respectively. It is observed that 180° phase shift occurs whenever Data1 changes as shown in Figure 8.14. Therefore, the polarity swap circuitry is controlled by Data1. On the other hand, "00" and "11" will generate a phase that corresponds to self-resonant frequency ω_{01}, whereas "01" and "10" will generate a phase that corresponds to self-resonant frequency ω_{02}. Therefore, XOR gate can be employed to switch between ω_{01} and ω_{02} as illustrated in Figure 8.15a. With this arrangement, QPSK modulation can be easily achieved. For O-QPSK modulation, Data1 is shifted by half symbol period with respect to Data0 before sending to the whole circuit.

An inverter-type output buffer is adopted in this design as shown in Figure 8.16. Its performance is limited by the QPSK modulation because of its nonconstant envelope nature. For O-QPSK with quasi-constant envelope nature, the linearity requirement is much relaxed. Therefore, nonlinear PA with high-power efficiency can be used (Liu et al. 2009). Fortunately, the targeted output power is limited to −3 dBm and the employed inverter-type PA can meet the linearity requirement with careful design (Sansen 2009). As shown in Figure 8.16, the PA includes a transimpedance stage and an inverter stage. R is around 50 kΩ, which settles the DC bias to around half of the supply voltage. From Feigin (2003), the linearity requirement of the PA can be worked out as follows:

$$P_{OIP3} = \frac{relative\_sideband + 3}{2} + P_{out}, \qquad (8.8)$$

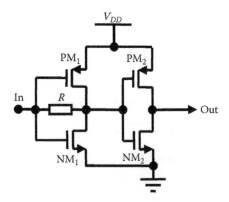

FIGURE 8.16 Inverter-type PA.

where P_{OIP3} is the desired output third-order intermodulation point in dBm (decibels relative to a milliwatt), P_{out} is the transmitter output power in dBm, and relative sideband is the relative t first side lobe suppression in decibels. With the targeted output power of -3 dBm and the first side lobe suppression of 12 dB (Ideal QPSK), the desired P_{OIP3} is 4.5 dBm. This leads to the output 1 dB compression point ($P_{out, 1 dB}$) requirement of -5.5 dBm.

In this design, the aspect ratios of PM_2 and NM_2 are chosen to be 25 µm/0.18 µm and 10 µm/0.18 µm, respectively. The PA output is connected to a 50-Ω load via a 3.5-pF AC coupling capacitor. It achieves a $P_{out, 1 dB}$ of -4.2 dBm with 20% power efficiency and the output power saturates at 1.13 dBm.

The measured EVMs are 5.97%/3.96% for QPSK/O-QPSK at a data rate of 50 Mbps/25 Mbps, respectively, as shown in Figure 8.17. Consuming 5.88 mW under

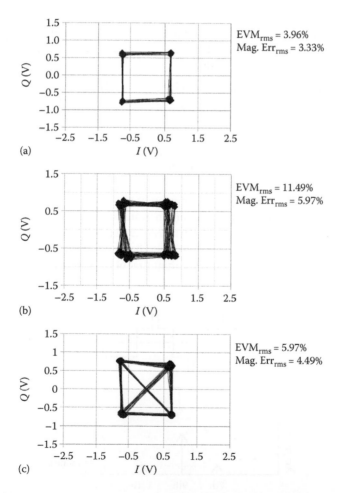

FIGURE 8.17 Measured EVM with 305 MHz injection signal: (a) O-QPSK at 25 Mbps, (b) O-QPSK at 50 Mbps, and (c) QPSK at 50 Mbps.

TABLE 8.1

Performance Summary

Injection Signal	1 V at 101.67 MHz	1 V at 305 MHz
Process	CMOS 0.18 μm	
Center frequency	915 MHz	
Supply voltage	1.4 V	
Die area	0.4 × 0.7 mm (active core)	
Modulation	QPSK/O-QPSK	
Maximum data rate	50 Mbps	100 Mbps
Power consumption (P_{DC})	5.88 mW	5.6 mW
	(PA: 2.24 mW	(PA: 2.24 mW
	ILO: 3.5 mW)	ILO: 3.22 mW)
Phase noise at 1 MHz	−119.4 dBc/Hz	−125 dBc/Hz
EVM	6.41% (QPSK)	5.97% (QPSK)/3.96% (O-QPSK)
Output power (P_{out})	−3.3 dBm	−3 dBm

1.4 V supply, it can achieve an energy efficiency of 118 pJ/bit while delivering an output power of −3 dBm. Its performance is summarized in Table 8.1.

For injection-locked LC oscillator, there is a trade-off between data rate and resulting spurs. To support higher data rate, a larger locking range is needed to speed up the setting time. However, this results in larger spurious tones because of limited harmonic suppression. Given the low-quality factor of on-chip inductors, the resulting spurious tones are in the range of −20 to −30 dBc and exceed the transmitter spectral requirement of <−40 dBc (Kavousian et al. 2008).

In Izad and Heng (2012), it has been demonstrated that the pulse shaping technique can be employed to suppress adjacent harmonics and result in spurious tone suppression of more than 22 dB. As the desired injected harmonic does not change, the resulting pulse shaping circuit is much simpler than that of Izad and Heng (2012b), and the

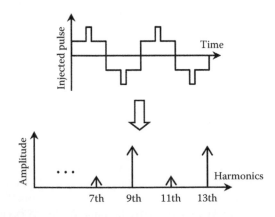

FIGURE 8.18 Pulse shaping with adjacent tone suppression.

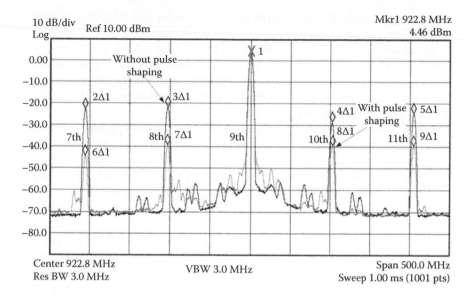

FIGURE 8.19   Spur reduction with and without pulse shaping.

idea is shown in Figure 8.18. The resulting spurious tone suppression with and without the proposed pulse shaping circuit is shown in Figure 8.19; as illustrated, more than 18 dB suppression has been achieved and the transmitter requirement can thus be met.

8.5.2 QPSK AND 8 PSK TRANSMITTER BASED ON ILRO

Figure 8.20 shows the block diagram of the proposed architecture. Inherent multiphases of a four-stage differential RO are directly employed to provide the PSK

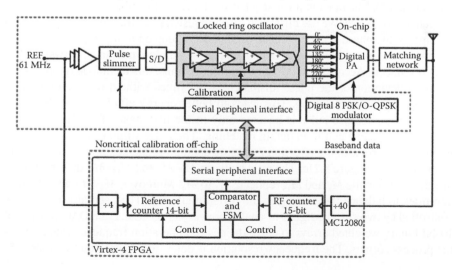

FIGURE 8.20   Proposed QPSK/8 PSK transmitter architecture.

modulation. Although RO suffers from poor phase noise and frequency instability, these problems are solved by subharmonic injection locking of the oscillator to the 15th harmonic of a 61-MHz *fundamental mode* crystal reference (Micro Crystal CC1F-T1A). Because an injection-locked oscillator behaves like a first-order PLL, phase noise and frequency stability are improved significantly. This architecture reduces the complexity of the multiphase carrier generation block, which leads to significant area and power saving.

Multiple phases of the carrier generated by the RO will drive a digital PA. Here, we take the advantage of digitally controlled PA to perform the modulation. A digital 8 PSK/O-QPSK modulator directly controls the PA to provide the modulated output to the antenna. Note that the proposed architecture is amenable to process and supply scaling because of its digital nature and can be easily ported to the newer technological nodes.

To compensate for process, voltage, and temperature variations, the free-running frequency of the oscillator is calibrated offline. A digital loop determines the code word required for digitally controlled delay elements in the RO. The frequency of the ring is digitally measured by counting the ring cycles for a fixed period. A successive-approximation search algorithm is then used to adjust the delay code. It is important to notice that this transmitter is intended to work in vivo and the temperature variations are small, which abates the need for frequent frequency calibration (Bae et al. 2011; Bohorquez et al. 2009). Thus, all the circuitries in this block can be turned off after calibration and the digital implementation enables digital storage of the control words with only leakage power. In this prototype, the calibration circuit is implemented off-chip using a field-programmable gate array for testing flexibility. The calibration engine can be easily ported on-chip. After synthesis and place and route, the calibration engine only occupies a negligible area of 35×45 µm.

As shown in Figure 8.21a, this prototype uses a four-stage current-starved pseudo-differential RO. The random mismatch between delay cells influences the accuracy of the generated phases. We use mismatch filtering resistors similar to Chen et al. (2010) to improve the phase accuracy. Since each node is coupled to the adjacent nodes by the resistors to average out the transition edges, any random mismatch between delay cells is reduced. According to Monte Carlo simulations, this technique improves the standard deviation of the phase mismatch by 4.7 times without the use of a larger transistor for better matching. This leads to better energy efficiency. Furthermore, although the reference pulse is injected only in the first stage of the RO, all other stages use the same cell to maintain matching. The pulse slimmer circuit shown in Figure 8.21b provides the injection pulses. A detailed implementation of each delay cell is shown in Figure 8.21c. The ratio between M_7–M_6 and M_{15}–M_{14} determines the injection ratio while M_8 and M_{13} replicate M_1 to M_4. Moreover, M_9–M_{12} are added as dummies to balance the loading in the differential path. The oscillator is digitally controlled by two current digital-to-analog converters (DACs). Each DAC consists of 10-bit binary weighted arrays to cover the desired operation frequency range across the process corners. The frequency resolution is determined based on Equation 8.5.

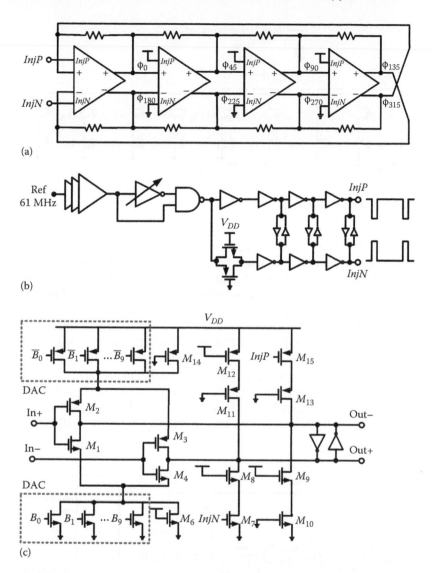

(a)

(b)

(c)

FIGURE 8.21 (a) Implementation of injection-locked RO. (b) Pulse generator. (c) Detailed schematic of the delay cell.

Figure 8.22 illustrates the digital PA with phase modulator circuit topology. It consists of eight unit amplifiers each comprising two transistors in series. The bottom transistors are driven by eight different phases of the RF carrier, tapped from the RO. The cascode transistors act as switches controlled by the digital baseband modulator. They select proper phases for each symbol. Note that each branch only amplifies a constant envelop (and constant phase) carrier while the (phase)

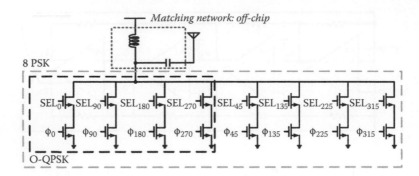

FIGURE 8.22 Power amplifier.

modulation is performed by switching between different branches. This allows for higher data rate modulation with much lower complexity compared to conventional PLL-based phase modulation and digital PA approach (Mehta et al. 2010). The combined output of all the amplifiers is connected to an off-chip matching network for impedance transformation to drive the antenna. One can either use all the unit amplifiers driven by an 8 PSK modulator (Figure 8.23) or only enable four of the unit amplifiers driven by a QPSK/O-QPSK modulator (Figure 8.23) to obtain a different modulation scheme.

The 8 PSK/O-QPSK transmitter is fabricated in a 65-nm digital CMOS (complementary metal-oxide semiconductor) with an active area of 0.038 mm^2. The RO can be tuned to oscillate from 758 to 950 MHz under 0.8 V supply. For the targeted output frequency of 915 MHz, the measured locking range of the oscillator extends

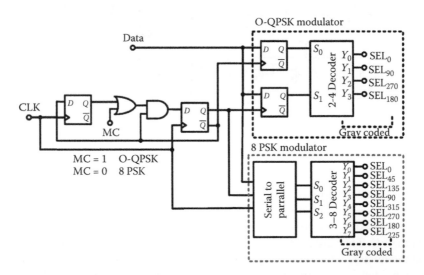

FIGURE 8.23 Digital baseband modulator.

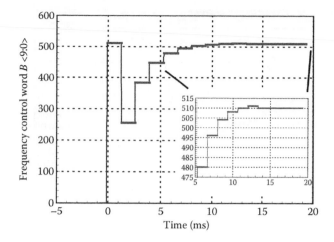

FIGURE 8.24   Measured convergence of the successive approximation algorithm.

from 902 to 927 MHz. Figure 8.24 shows the measured convergence of the successive approximation algorithm, which requires only 10 cycles to converge.

The measured phase noise of the RO under locked and unlocked conditions is shown in Figure 8.25a. As illustrated, injection locking significantly improves the phase noise from 1 kHz to 1 MHz offset. The total measured integrated RMS jitter for the locked RO is 4.6 ps (1.52°) while the worst-case reference spurs are less than −47 dBc (Figure 8.25b). The measured settling time of the RO is shorter than 80 ns, which allows aggressive duty-cycled operation. Figure 8.25c shows the output spectrum for 8 PSK mode at 55 Mbps. As illustrated, this prototype meets the Federal Communications Commission spectral mask and thus the interference to other wireless standard would not be a concern.

Figure 8.26 shows the measured constellation for the 8 PSK and O-QPSK mode. A similar measurement has been carried out on 10 chips to investigate the effect of mismatch. The worst-case measured EVMs at a data rate of 55 Mbps are 3.8% and 4.46%, respectively. A typical 8 PSK/O-QPSK receiver requires an EVM of less than 15/23% to achieve a bit error rate better than 10^{-4}. The measured EVM in our transmitter meets this requirement with a good margin.

The RO consumes 538 μW and the PA dissipates 286 μW while delivering −15 dBm output power. The delivered power has included loss of PCB and matching network. The total power consumption (including crystal oscillator) in any mode of operation is less than 938 μW.

Table 8.2 summarizes the performance of the prototype and compares it with the state-of-the-art designs. This work achieves the energy efficiency of 17 pJ/bit, which is 8× smaller than that of the state of the art. It also has the lowest area without requiring any off-chip inductors (other than matching network). This prototype is the first transmitter that provides spectral efficient 8 PSK modulation with power consumption in the sub-milliwatt range, confirming the feasibility of achieving higher data rates with low-power consumption.

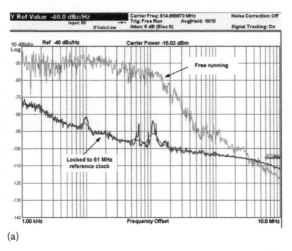

(a)

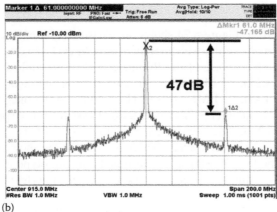

(b)

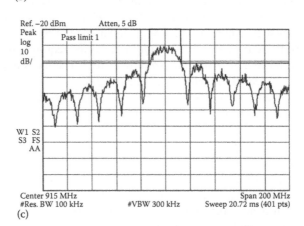

(c)

FIGURE 8.25 (a) Measured phase noise of the RO under locked and unlocked conditions. (b) Measured unmodulated output spectrum of the carrier at −15 dBm output power. (c) Measured output spectrum with output power of −15 dBm at 55 Mbps.

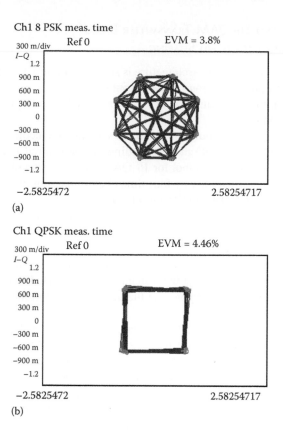

FIGURE 8.26 Measured constellation at (a) 8 PSK and (b) O-QPSK at 55 Mbps.

TABLE 8.2
TX Performance Summary and Comparison

	Daly JSSC Dec 07	Vidojkovic ISSCC Feb 11	Liu CICC Sep 08	Bae JSSC Apr 11	This Work
Frequency (MHz)	900	2400	400	920	915
Modulation	OOK	OOK	O-QPSK	FSK	8 PSK or O-QPSK
Data rate (Mbps)	1	10	15	5	55
Output power (dBm)	−11.4	0	−15	−10	−15
Power diss. (mW)	3.8	2.53	3.48	0.7	0.938
Active area (mm²)	0.27	0.882	0.7	1.65[a]	0.038
Energy/bit (nJ/bit)	3.8	0.253	0.23	0.14	0.017
Supply voltage (V)	0.8–1.4	1	1.2	0.7	0.8
Technology	0.18 μm	90 nm	0.18 μm	0.18 μm	65 nm

[a] Chip area. It uses four off-chip inductors.

8.5.3 QPSK AND 16-QAM TRANSMITTER BASED ON ILRO WITH BAND SHAPING

The proposed transmitter (TX) architecture is shown in Figure 8.27. ILRO forms the core of TX, which provides a four-phase output (φ_0, φ_{90}, φ_{180}, φ_{270}) with good phase noise. Direct quadrature modulation at PA is proposed here to provide both phase and amplitude modulations.

To achieve the desired modulation and band shaping, the incoming serial data are first converted into parallel I/Q data depending on the desired modulation (2 bits/symbol for QPSK and 4 bits/symbol for 16-QAM). If band shaping is activated, the I/Q data will then be upsampled by four times before passing through a root raised cosine (RRC) filter ($\alpha = 0.4$). Finite impulse response filters instead of read-only memory (ROM) are adopted here for RRC filter implementation to provide flexibility in filter coefficient tuning. After that, it is further upsampled by two times before going through an interpolation filter. The upsampling will push the unwanted image further away from the targeted output and can thus be suppressed easily by matching network and antenna. If band shaping is deactivated, the I/Q data will be sent to the decoder right away, bypassing the intermediate upsampler and RRC filter to save energy.

In this implementation, an injected reference of 100 MHz is chosen and the ILRO will lock to the ninth harmonic of the injected reference. Similar to earlier work, a frequency calibration algorithm has been incorporated on-chip to fine-tune the RO free-running frequency to match the ninth harmonics of the injected reference. Unlike Izad and Heng (2012a), which use an off-chip reference source, the fundamental mode 100-MHz crystal oscillator is also built on-chip to better evaluate the phase noise and energy efficiency performance of the TX.

An identical four-stage pseudodifferential RO shown in Figure 8.21 with mismatch filtering technique is employed. This helps improve the energy efficiency of the RO. The digital PA with embedded phase multiplexer and amplitude control is shown in Figure 8.28. It consists of four amplifier cores driven by four output phases

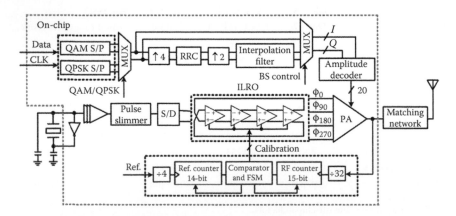

FIGURE 8.27 Proposed QPSK/16-QAM TX architecture.

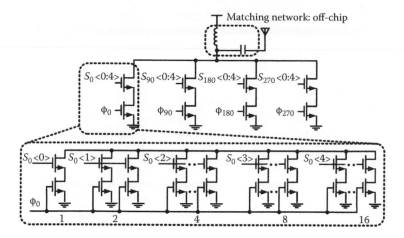

FIGURE 8.28   Digital PA with direct phase and amplitude modulation.

(φ_0, φ_{90}, φ_{180}, φ_{270}) from ILRO. The bottom transistors are connected to respective ILRO output phases, whereas the top transistors are used to activate the corresponding phase. To achieve 5-bit amplitude control for each output phase, the transistors within each amplifier core are further segmented into an array of 31 transistor pairs. Direct quadrature modulation is achieved through current summing from two activated phase branches with different current amplitudes. The combined output current is then sent to an off-chip impedance matching network before driving the 50-Ω antenna. The 20-bit control (5-bit/phase $\times$ 4 phases) for the PA is provided from an amplitude decoder.

Fabricated in 65-nm CMOS, the TX occupies an active area of 0.08 mm^2. The only off-chip components needed are a matching network and a 100-MHz crystal. The measured tuning range of RO covers from 0.81 to 1 GHz. The measured locking range of the ILRO is from 885 to 925 MHz.

Figure 8.29 shows the measured phase noise under injection locking. By setting the crystal oscillator power consumption to 115 µW, the ILRO achieves a total integrated RMS jitter of 1.54°. If the crystal oscillator power consumption is allowed to increase to 380 µW, 7 dB improvement in phase noise is noted. The ILRO also shows a fast startup time of less than 88 ns in Figure 8.30, which is critical for burst-mode operation. With frequency calibration, the reference spur of the ILRO can be lowered to −56 dBc as shown in Figure 8.31.

The measured constellation for QPSK/16-QAM with and without BS is shown in Figure 8.32. EVMs better than 6% are observed for QPSK at 50 Mbps and 16-QAM at 100 Mbps. Figure 8.33 plots the EVM performance at different data rates under different modulations. Only 1% EVM variation is observed for the collected data over 10 chips, showing the robustness of the proposed TX.

Figure 8.34 presents the output spectrum with a fixed data rate of 50 Mbps. As illustrated, 38 dB side lobe suppression is achieved with BS, which is 25 dB more compared to TX without BS. In addition, the 16-QAM mode is also twice more spectral efficient than QPSK.

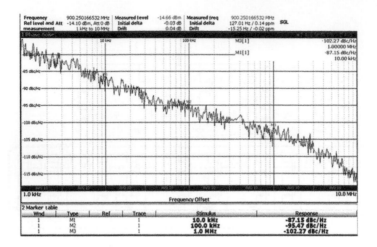

FIGURE 8.29   Measured phase noise under injection locking.

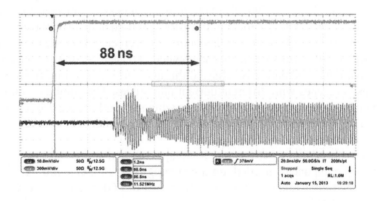

FIGURE 8.30   Measured settling time.

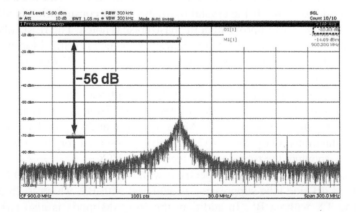

FIGURE 8.31   Measured spurious tone performance of ILRO.

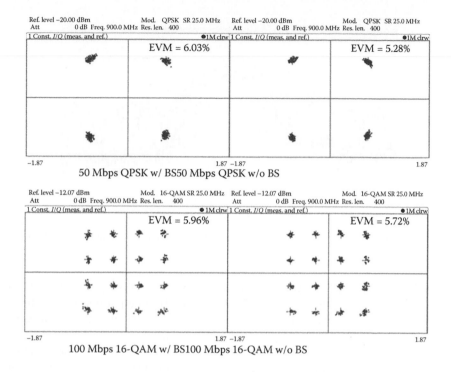

50 Mbps QPSK w/ BS   50 Mbps QPSK w/o BS

100 Mbps 16-QAM w/ BS   100 Mbps 16-QAM w/o BS

FIGURE 8.32   Measured EVM for QPSK and 16-QAM at 25 Mbps with and without band shaping.

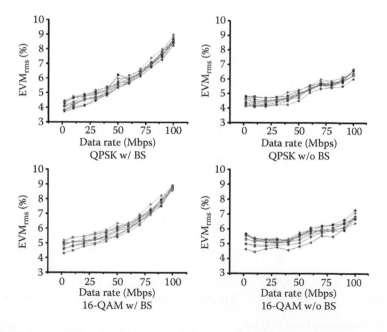

FIGURE 8.33   Measured TX EVM variations versus data rate across 10 chips.

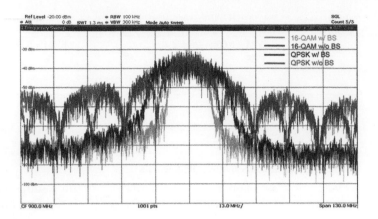

FIGURE 8.34 Comparison of output spectrum with/without band shaping for QPSK and 16-QAM at 50 Mbps.

Under a 0.77-V supply, the TX consumes 2.6 and 1.3 mW, respectively, with and without BS while transmitting at 25 Msps with −15 dBm output power. The digital portion that provides band shaping consumes 50% of the total power. This power can be further reduced by adopting an ROM-based RRC filter.

The TX performance is compared with other similar multi-PSK and 16-QAM TX in Table 8.3. We achieve the highest data rate of 100 Mbps and energy efficiency of 13 pJ/bit (without band shaping) compared with others. Because of the

TABLE 8.3

PSK/QAM TX Performance Summary and Comparison

References	Liu ASSCC'11	Diao ASSCC'10	Izad CICC'12	Zhang RFIC'12	This Work
Frequency (MHz)	2400	900	915	350–578	900
Modulation	HS-OQPSK	O-QPSK/QPSK	O-QPSK/8PSK	16-QAM[a]	QPSK/16-QAM
Data rate (Mbps)	2	50	55	7.5	50/100
Output power (dBm)	−3	−3.3/−15	−15	0.23	−15
Power (mW)	15	5.88/3	0.938	4.9	1.3 (w/o BS) 2.6 (w/BS)
Area (mm²)	0.35	0.28	0.038	0.7	0.08
Band shaping	Yes, >29 dB	No	No	No	Yes, >38 dB
Energy/bit (nJ/bit)	7.5	0.12/0.06	0.017	0.65	0.026/0.013 (w/o BS) 0.052/0.026 (w/BS)
Supply (V)	1.5	1.4	0.8	1.5	0.77
Technology	0.18 μm	0.18 μm	65 nm	0.18 μm	65 nm

[a] Circular constellation-based 16-QAM.

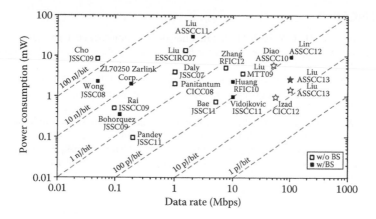

FIGURE 8.35   Energy efficiency comparison of low-power transmitters.

simplicity of our TX architecture, the energy efficiency only worsens to 26 pJ/bit with band shaping.

8.5.4 COMPARISON

As a comparison, we have plotted all low-power transmitters with and without band shaping in Figure 8.35. The *x*-axis indicates the achievable data rate while the *y*-axis indicates the power consumption of the corresponding transmitter. The dashed diagonal lines imply constant energy efficiency boundaries. The transmitter with better energy efficiency will be approaching the bottom-right corner. The energy efficiency of the three fabricated transmitters discussed earlier has been plotted. As illustrated, the three designs reported the lowest energy efficiency for both transmitters with and without band shaping, clearly showing the advantage of the proposed techniques.

8.6 CONCLUSION

In this chapter, we have reviewed the need for an energy-efficient high–data rate transmitter. We have also discussed the trend of moving toward complex modulation to maximize spectral efficiency. Through the proposed techniques, such as ILO, ILRO, and ILRO with direct quadrature modulation, energy-efficient PSK and QAM transmitters can be achieved, which report one order improvement on energy efficiency, reaching 13 pJ/bit.

ACKNOWLEDGMENTS

We thank MediaTek for sponsoring the fabrication. We would also like to thank the Institute of Microelectronics for facilitating the packaging and testing of some of the works. Last but not the least, we would like to thank Shengxi Diao, San Jeow Cheng, Mehran Mohammadi Izad, and Xiayun Liu for the designing and testing of the chips.

REFERENCES

Bae J., L. Yan, and H. J. Yoo. 2011. A low energy injection-locked FSK transceiver with frequency-to-amplitude conversion for body sensor applications. *IEEE J. Solid-State Circ.* 46 (4): 928–937.

Bohorquez J., A. P. Chandrakasan, and J. L. Daeson. 2009. A 350 µW CMOS MSK transmitter and 400 µW OOK super-regenerative receiver for medical implant communications. *IEEE J. Solid-State Circ.* 44 (4): 1248–1259.

Chen M. S. W., D. Su, and S. Mehta. 2010. A calibration-free 800 MHz fractional-N digital PLL with embedded TDC. *IEEE J. Solid-State Circ.* 45 (12): 2819–2827.

Diao S., Y. Zheng, Y. Gao et al. 2012. A 50-Mb/s CMOS QPSK/O-QPSK transmitter employing injection locking for direct modulation. *IEEE Trans. Microw. Theory Tech.* 60 (1): 120–130.

Feigin J. 2003. Don't let linearity squeeze with accurate models, the range- and data-rate-limiting. *Commun. Systems Design*, 12–16.

Izad M. M. and C. H. Heng. 2012a. A 17 pJ/bit 915 MHz 8 PSK/O-QPSK transmitter for high data rate biomedical applications. *Proc. CICC*, 1–4. San Josè, USA, September 09–12.

Izad M. M. and C. H. Heng. 2012b. A pulse shaping technique for spur suppression in injection-locked synthesizers. *IEEE J. Solid-State Circ.* 47 (3): 652–664.

Kavousian A., D. K. Su, M. Hekmat et al. 2008. A digitally modulated polar CMOS power amplifier with a 20-MHz channel bandwidth. *IEEE J. Solid-State Circ.* 43 (10): 2251–2258.

Kim K., S. Lee, E. Cho, J. Choi, and S. Nam. 2010. Design of OOK system for wireless capsule endoscopy. *Proc. ISCAS*, 1205–1208. Paris, France, May 30–June 2.

Liu Y.H., C.L. Li, and T. H. Lin. 2009. A 200-pJ/b MUX-based RF transmitter for implantable multichannel neural recording. *IEEE Trans. Microw. Theory Tech.* 57 (10): 2533–2541.

Mehta J., R. B. Staszewski, O. Eliezer et al. 2010. A 0.8 mm all-digital SAW-less polar transmitter in 65 nm EDGE SoC. *ISSCC Dig. Tech. Papers*, 58–59. San Francisco, USA, February 7–11.

Razavi B. 2004. A study of injection locking and pulling in oscillators. *IEEE J. Solid-State Circ.* 39 (9): 1415–1424.

Sansen W. 2009. Low-power, low-voltage Opamp design. ISSCC Short Course.

9 Toward BCIs Out of the Lab

Impact of Motion Artifacts on Brain–Computer Interface Performance

Ana Matran-Fernandez, Davide Valeriani, and Riccardo Poli

CONTENTS

9.1 INTRODUCTION

9.1.1 BRAIN–COMPUTER INTERFACES

A brain–computer interface (BCI) is a system that provides a direct pathway between the brain and an external device, allowing people to act on the world without moving any muscle. In order to do so, the BCI needs access to the central nervous system and, in particular, to the brain. Invasive BCI systems are those that are implanted directly into the brain of the user [1,2]. They allow the detection of cleaner signals, but they need surgery to be installed. On the other hand, in noninvasive BCIs, the insight into the neural processes is most frequently given by electroencephalography (EEG) systems that record the electric activity of the brain from the scalp of the user [3–15]. EEG-based BCIs have several advantages: they are easy to use, are compact, and provide a very good time resolution. Thanks to these, several applications have been suggested in the last decades for EEG-based BCIs.

Initially EEG-based BCIs were developed as tools to enhance the quality of life of people with severe motor disabilities (e.g., locked-in) [12,16]. In this application, the main focus is to provide systems that are reliable, efficient, and easy to use in a real context, which includes their portability. They have also been used as passive tools for able-bodied users (e.g., for monitoring driver's attention levels in a car or as an extra input to control a videogame) [17,18]. Moreover, if the EEG recording system is discreet enough, portable BCIs could be conceived as a wearable device, for monitoring changes in brain activity through the day [19]. Indeed, EEG recordings are also routinely used for medical applications, for example, for the characterization of epileptic seizures and study of sleep and brain lesions [20–22].

While until very recently all BCIs involved an individual user, collaborative BCIs (i.e., BCIs that are used simultaneously by several users to control one device) have now started to show their advantages in speed and accuracy with respect to non-BCI users and single-user BCIs. There are indications that, in the future, they could be adopted for use in the workplace, for example, by intelligence analysts [23,24] or in group decision making [25].

9.1.2 ARTIFACTS AND THEIR EFFECTS ON BCIs

Despite all the advantages of these noninvasive BCIs, they also present a major drawback associated with the nature of EEG: the neural signals recorded from the scalp are highly affected by a variety of artifacts [26,27]. These include the electrical activity of the heart (electrocardiogram), inconsistent contact of the electrodes, (for example, because of movements), and muscle artifacts (electromyogram [EMG]) such as those induced by neck movements, ocular activity (electrooculogram [EOG]), eye blinks, swallowing, and so on. These artifacts can affect the quality of the EEG signals recorded, consequently deteriorating the performance of the BCI system.

EOG artifacts, for instance, may be orders of magnitude bigger than the ordinary elements of EEG. Therefore, when they are present in the recording, it is usually impossible to detect the small variations of the voltages that represent the brain processes. EOG artifacts can also lead to extremely larger deviations from the desired behavior in BCIs, particularly those where there is an analog relationship between input signals and output control signals, such as in the Essex BCI Mouse [15,28,29].

For these reasons, the control of artifacts is a very important step in BCI research.

9.1.3 ARTIFACT CORRECTION AND REJECTION AND THEIR LIMITATIONS

Fundamentally, there are two approaches one can take to deal with artifacts: *correction* and *rejection* [27,30].

Correction means subtracting out the effects of the artifacts from the EEG signal. This requires estimating the contribution of the artifacts on the recorded signal. For example, the effects of eye blinks and vertical components of saccades can be reduced by using the time-domain linear regression between each channel and the vertical EOG [27,31], by using dipole localization procedures [32], or by applying independent component analysis (ICA) [33] to decompose EEG signals into independent components (sources variability) and subtract from the EEG those that represent pure artifacts [34–38]. EMG artifacts could be removed by applying a low-pass filter to the EEG signal or by using ICA or regression methods [30,39].

However, while the correction technique is well used in literature to deal with EOG artifacts, its application to other sources of noise is not straightforward. Many more types of artifacts are not only possible but ubiquitous if the EEG is recorded outside the laboratory (for example, for utilization in a portable BCI). Under laboratory conditions, experimenters usually ask the volunteers to hand over their mobile phones to avoid noise from electrical devices, and the instructions given in the protocol ask them to remain as still as possible, try to avoid eye blinks, and so on. While these measures are widely accepted in event-related potential (ERP) studies, they are not realistic for the daily use of a BCI.

Moreover, despite the wide use of artifact correction methods, some skepticism remains as to whether these are actually introducing distortions in the corrected EEG data [32]. In order to validate an artifact correction method, one would have to first contaminate data and then show how the corrected data compare to the original signal before contamination.

The second approach is rejection, where fragments of EEG (or trials) affected by artifacts are simply removed from the analysis. Rejection requires the *detection* of artifacts (while correction does not necessarily require it). For instance, two of the simplest and fastest automated detectors for EOG artifacts are as follows [27,40]: (a) deleting portions of the data where the EOG amplitude deviates by more than a set threshold, say, 50 μV or 75 μV, from the baseline; (b) measuring the difference, ΔV, between the maximum and minimum EOG amplitude within a block of the signal and rejecting it if ΔV exceeds a threshold.

Artifact detectors usually require positioning some extra electrodes on the subject (e.g., around the eyes for EOG detection), making the BCI system less portable. Although there are a number of effective techniques to detect and remove artifacts (e.g., see Refs. [38,41] for EOG) without the availability of these extra electrodes, they are quite complex, especially considering that they should be applied for various types of artifacts that could affect the EEG signal.

Beyond the system's complexity, the main disadvantage of artifact rejection is that it can significantly reduce the amount of epochs available for ERP analysis [32]. In the BCI field, discarding epochs that contain artifacts would slow down the performance of the BCI. Depending on the percentage of discarded trials, this might not be acceptable for the user. For this reason, since various artifacts can affect the same EEG signal, not all contaminated trials can be rejected and, therefore, the rejection approach is usually not sufficient to get "clean" signals.

Therefore, it is unlikely to be able to rely on EEG signals that are not affected by noise in an EEG-based portable BCI system. If we first focus on the utilization of a portable BCI by a disabled person, even in the case of he or she being completely locked-in, the environment will introduce noise in the EEG signals that should be processed by the system. In this extreme case, it is important that the BCI is built so that its performance is not significantly affected by such interference. In the case of able-bodied users, the main sources of noise will probably be motion artifacts, and it is thus important to know which types of artifacts will be more likely to affect performance and how robust a BCI can be with respect to them.

9.1.4 CONTRIBUTIONS OF THIS STUDY

To start assessing how close we are to performing a transition from the laboratory (with seated and static users and noise- and distraction-free environments) to the real world (which is noisy, dynamic, and distracting, and in which users constantly move and perform other actions besides controlling a computer with their brains), in this chapter we studied the impact of several types of artifacts on the performance of a BCI. In particular, we wanted to see whether the trained classifier that virtually all modern BCIs include could cope with the artifacts.

More specifically, in the following, we will analyze the impact of more than 20 different types of artifacts on a portable EEG-based BCI system. Instead of dealing with artifact rejection or correction, considering the limitations described above, our aim is to investigate to what degree a portable BCI applied to everyday life can still adequately perform in comparison to a BCI used under laboratory conditions and in the absence of noise and artifacts. By considering this high number of different types of artifacts, we may be able to figure out which of those have a high impact on the performance of the portable BCI system and, on the other hand, for which ones the system is robust enough to ignore them. Relatively little is known about how much the performance of a portable BCI would be affected by such a wide range of artifacts.

The analysis will focus on the artifacts that reflect movements that are typical in a driving scenario. Apart from the "classical" artifacts (like eye movements and eye blinks), actions like moving the head or changing the gear with one hand will also be included. We chose a driving scenario since many possible applications of BCI systems are in the automotive area [42]. The Essex BCI Mouse has been shown to be suitable for control of a spaceship in a videogame [43]. It is possible to envision a system in which a similar controller can be embedded on the windscreen of a car, in a way that allows for brain control while driving. Also, portable BCI systems could be integrated in a car to monitor the level of attention of the driver and raise an alarm in case of danger, in order to reduce the possibility of accidents. However, one of the main limitations of this application is precisely the large amount of noise generated by the movements performed by the driver while he or she is driving. This is the scenario on which we focus for the rest of the chapter.

While this serves as a good framework for a first approach to the topic of portable BCI, it has obvious limitations. For example, we do not study here walking-derived artifacts, which of course cannot be applied in a driving task. However, the conclusions of this study can be applied not only to driving scenarios but also for wheelchair users, gamers, and, in general, any other BCI application in which the able-bodied user of the BCI performs a task while remaining seated.

Moreover, this study contributes to bringing BCI out of the laboratory. The conclusions drawn about the impact of different types of artifacts could be used for portable BCI applications, such as wearable devices, to reduce the complexity of the system, focusing the processing mechanisms for artifact correction on those types that affect BCI performance most severely.

The chapter has the following structure. Section 9.2 describes the protocol we used to record the neural data in the original BCI protocol and the data used to add different artifacts to it. Section 9.3 describes the methodology applied to process, classify, and analyze the data. In Section 9.4, we show and discuss the results obtained with the clean and the artifact-contaminated data sets. Finally, in Section 9.5 we draw some conclusions about this work and include suggestions for further work.

9.2   DATA COLLECTION

This section will focus on describing how the neural data were collected. As we mentioned in Section 9.1, rather than utilizing data from a portable BCI, we artificially added artifacts to BCI data collected under laboratory conditions in previous experiments and studied the effects of this added noise on BCI performance. We will begin by describing the BCI data (i.e., data acquired during a real BCI experiment) and will then continue with an explanation of the protocol that we followed to collect new data containing different types of artifacts.

In both cases, data were acquired with a BioSemi ActiveTwo EEG system. Neural data were recorded from 64 electrode sites (see Figure 9.4), organized according to the 10–20 system, and referenced to the mean of the electrodes placed on both earlobes.

9.2.1   BCI DATA

In the present study, we used a subset of the data gathered from 16 participants (average age of 30, all with normal or corrected to normal vision except for one who had strabismus with exotopia in the left eye) in the two experiments reported in Ref. [14], where a variety of visual stimulation protocols to be used in a BCI mouse were tested.

As illustrated in Figure 9.1, in the experiments, participants were presented with a display containing eight circles (with each circle representing a direction of movement for the mouse cursor) that formed an imaginary circle at the center of the display. The circles (stimuli) flashed sequentially (by changing rapidly from a baseline color to a different color and back) for 100 ms each without any delays in between.

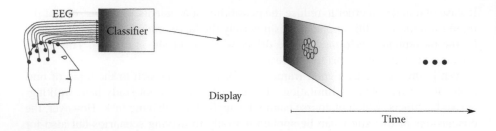

FIGURE 9.1   Stimulation protocol used in our BCI mouse.

This meant that all eight different stimuli from the imaginary circle flashed within 800 ms, forming what we call a *circle epoch*.

The experiments were divided into runs, which we call *direction epochs*. Each direction epoch contained between 20 and 24 circle epochs and, thus, lasted between 16 and 19.2 s.

At the beginning of each direction epoch, participants were assigned a target circle and asked to perform the task of mentally naming its color every time it flashed. As we will detail below, the flashing of a circle constitutes the beginning of a *trial epoch*. If the flashed circle was the target circle, the trial epoch was labeled as a *target*. Otherwise, it was labeled as a *nontarget*.

Each participant carried out 16 direction epochs, with each circle being a target in two of them.

9.2.2 ARTIFACTS

As we indicated above, in this chapter, we will be focusing on BCIs that are controlled while the user is driving a car. Hence, we chose a pool of possible actions that are typically performed while driving. This activity involves doing several different actions, such as looking at side mirrors or changing the gear, that could generate artifacts in the EEG recording.

More specifically we decomposed it into 24 different activities representing eye, face, neck, arm, and leg movements. The complete list comprises turn/bend neck to the left/right, move head up/down, move eyes up/down/to the left/to the right, move tongue with the mouth closed, blink once/repeatedly, turn the wheel left/right, change gear with left/right hand, yawn, swallow, count in loud voice, cross left (respectively, right) foot over right (respectively, left), squint, and a baseline "do nothing" condition. Some of these activities have been depicted in Figure 9.2.

To generate artifactual EEG signals, we asked a volunteer (27-year-old female) to perform the 24 actions (one at a time) while we were recording her brain activity. She was asked to perform 15 repetitions of each action, for a total of 360 trials. The order in which these were performed was randomized.

The protocol used was as follows. The participant was presented with a display showing a black screen. The screen then showed the name of the task (written as in Figure 9.2). To avoid ERPs and neural activity related to the processing of the task, the participant was asked to wait for 1 s after the onset of the task before performing

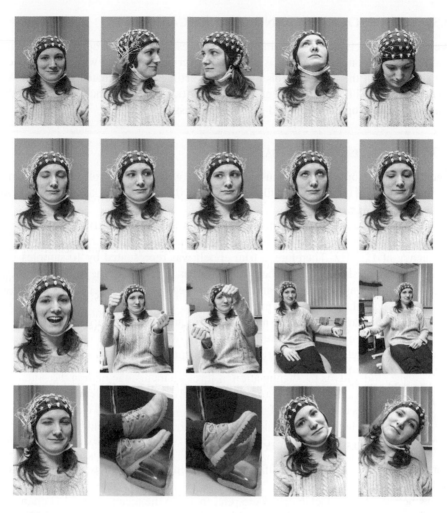

FIGURE 9.2 Movements performed to generate noise. From top left to bottom right: do nothing, turn neck to the left, turn neck to the right, move head up, move head down, blink, move eyes to the left, move eyes to the right, look up, look down, yawn, turn the wheel left, turn the wheel right, change gear with the left hand, change gear with the right hand, squint, cross right foot over left, cross left foot over right, bend neck to the right, bend neck to the left.

the required action. She then had a 3-s time window to perform the action. Five seconds after the onset of task presentation, the screen turned black again and stayed static until the participant indicated, through the press of a mouse button, that she was ready to move for the next task.

The volunteer was seated on an electrically adjustable chair that had been unplugged from the mains to prevent it from injecting electromagnetic noise in the EEG data. After the recording of the 24 types of movements, she was asked to remain still (the screen showed the "do nothing" instruction) while the chair was reconnected to the mains. An extra 15 trials were recorded in this scenario.

9.3 DATA PROCESSING AND CLASSIFICATION

We begin this section by describing the preprocessing stage for both the BCI data and the epochs containing artifacts. We then describe the way in which we blended the signals from the two protocols to generate artifact-contaminated BCI data. We end the section with a description of our classification approach, including our methods for electrode optimization and performance assessment.

9.3.1 PREPROCESSING BCI DATA

The EEG data from the BCI mouse experiment were initially collected at a sampling rate of 2048 Hz, band-pass filtered between 0.15 and 40 Hz, and finally downsampled to 512 Hz.

From each channel (or electrode), an 800-ms trial epoch starting at the flashing of a circle was extracted and further decimated (with averaging) to a sampling frequency of 12.5 Hz. Therefore, a trial epoch was represented by 10 samples per channel. If the flashed circle had been identified as the target of the direction epoch, that trial epoch was labeled as a target epoch, whereas nontarget epochs were those that started with the flashing of a nontarget circle.

Since the nontarget trials were much more frequent than the target ones (7 out of every 8 trials were nontargets), we balanced the two classes by randomly subsampling the nontarget trials on a participant-by-participant basis, in order to improve the performance of the classifier.

The data sets created from these data will be referred to in the rest of the chapter as D_0.

9.3.2 PREPROCESSING ARTIFACTS

For each trial, we extracted an artifact epoch starting 1 s after the presentation of the task and lasting 3 s.

As before, also artifact epochs were referenced to the mean of all EEG channels and to the mean of the earlobes, band-pass filtered between 0.15 and 40 Hz and downsampled by a factor of 4, to the final sampling rate of 512 Hz.

Since these artifacts were added to "clean" BCI data from a different experiment, we then detrended each trial to ensure that the beginning and end of the epochs were 0 and, thus, no discontinuities were introduced in the BCI data when adding an artifact.

One to three detrended artifact epochs were then added to the BCI data at random positions in each direction epoch (more on this in Section 9.3.3).

9.3.3 CREATION OF SIMULATED PORTABLE BCI DATA SETS

In order to determine the impact of different types of artifacts on a BCI, we created a number of simulated portable BCI data sets, each of which consisted of "noisy" epochs $Ep_{i,j,k}$ for each participant of the BCI mouse experiment, where $i \in \{1,\dots,16\}$ is the direction epoch from which the BCI data are extracted, $j \in \{1,\dots,25\}$ is the

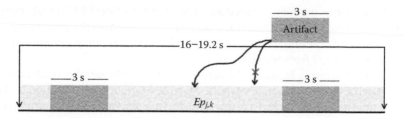

FIGURE 9.3   Process of creation of $Ep_{j,k}$.

type of artifact that has been added to the neural data, and $k \in \{1,2,3\}$ is the number of occurrences of that type of artifact in each direction epoch.*

Each channel of $Ep_{i,j,k}$ is obtained by adding a direction epoch extracted from the BCI data (Ep_i) and a new epoch, which we call $Ep_{j,k}$, of the same length as Ep_i, which contains k artifact epochs from the new signals recorded from our volunteer. That is,

$$Ep_{i,j,k} = Ep_i + Ep_{j,k}. \tag{9.1}$$

The process of creation of $Ep_{j,k}$ is depicted in Figure 9.3. First of all, a new structure of the same length of Ep_i is created and filled with 0's. Second, k random artifact epochs from the pool of 15 of type j are selected. These are added to the newly generated structure at random locations with the condition that no more than one artifact could be present at each sample.

In order to have data that are consistent, each channel of the recorded artifact was added to the corresponding channel of the BCI data. That is, each generated channel of $Ep_{i,j,k}$ contains the summation of BCI data from that channel and k randomly placed artifacts as read on the same electrode site.

Finally, after the creation of $Ep_{j,k}$, this structure is added to a given run of the BCI mouse data (Ep_i) before extracting and preprocessing the individual 800-ms trial epochs that are used for classifying trials.

Since we add between one and three 3-s artifact epochs to direction epochs lasting between 16 and 19.2 s, the percentage of contamination in the data sets ranges from 15%–19% (depending on the length of a direction epoch) for $k = 1$ to 47%–56% for $k = 3$.

The procedure described above creates a total of 75 data sets for the simulated portable BCI. We will term each of these $D_{j,k}$, where, as above, j is the type of artifact that has been used to contaminate the original BCI data set and k is the number of artifact epochs added to a direction epoch.

9.3.4   FEATURE EXTRACTION

Various features have been used in BCI research to extract meaningful information from the EEG signal and improve the classification performance. In order to capture as

---

* Since the epochs extracted and preprocessed from the artifact experiment (3 s) are much shorter than those of a direction epoch of the BCI mouse (Ep_i, between 16 and 19.2 s as described in Section 9.2.1), it is possible to have multiple artifacts occurring within a run. This allowed us to investigate the effect of different levels of contamination in the BCI.

much information as possible, we used an *optimal subset* of the EEG signals recorded from the 64 electrode sites as neural features that were given to the classifiers.

The process of choosing the best subset of channels was composed of two steps: *feature reduction* and *feature selection*.

First, we performed a feature reduction step to reduce the number of electrodes to be considered for the optimization process. We randomly split the original BCI data set into a training set (80%) and a test set (20%), keeping the same proportion of target and nontarget trials as in the original data set D_0. Then, we normalized both sets by subtracting the mean value and dividing by the standard deviation of the training set. The test set was kept separate for later use (described in Section 9.3.5) while we used a five-fold cross-validation loop on the training set to train and validate a Fisher discriminant analysis (FDA) classifier that used all possible combinations of subsets of electrodes as features. We then used the area under the curve (AUC) of the receiver operator characteristics (ROC) to rank the different subsets. The AUC is a well-known summary for ROC curves that has been used widely in machine learning—more on this in Section 9.3.5. By looking at each best subset of electrodes of each participant, we manually selected the electrodes that appeared in the best set at least in seven out of the eight volunteers. These were the electrode sites O2, PO8, Cz, PO4, PO7, PO3, Pz, Oz, and P8 shown in Figure 9.4. These electrodes cover the area where the P300 ERP is most easily detected.

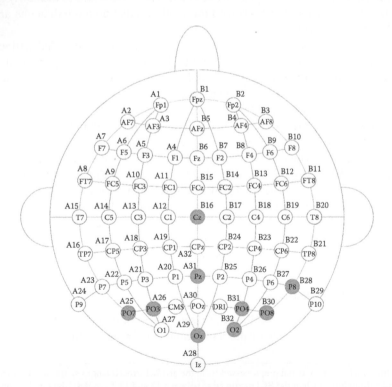

FIGURE 9.4 The electrodes available in our EEG acquisition system. The gray locations represent the electrodes selected after feature reduction.

Second, we used this optimal set as a pool of electrodes in the *feature selection* step that was common for all participants. Then, using the training set from either the original D_0 data set or the simulated portable BCI data sets, we tested all the possible subsets of between three and nine electrodes to find the best combination of features for each participant and type of data set using a five-fold cross-validation loop. The features from the optimal sets of electrodes found in this way were then used to train individually tailored classifiers, as we will explain in detail in Section 9.3.5.

9.3.5 CLASSIFICATION

In order to classify the features extracted from D_0 and the artifact-contaminated data sets into the target and nontarget classes, we relied again on the FDA classifier, since it is frequently used in BCI research and it is efficient for real-time portable BCI use.

First, for each participant ($p = 1,...,8$), we used the optimal subset of electrodes that had been derived as described in Section 9.3.4 to extract features from the training set of D_0. These were used to train a classifier C_p that was specific for each participant, as is commonly done in BCI.

Then, we used the trained C_p to predict the classes (target versus nontarget) of the unseen trial epochs from the test sets of D_0 and $D_{j,k}$ (the original BCI data set and the simulated portable BCI data sets, respectively). Results from this first experiment are reported in Sections 9.4.1 and 9.4.2.

After training, the output of the FDA classifier can be interpreted as a measure of how closely the feature vector associated with the stimulus matches the target. By applying a threshold to this measure, one can transform it into a binary decision regarding the presence of a target. Naturally, the higher the threshold is, the less likely a false-positive error will be. However, unavoidably, a higher specificity brings a lower sensitivity (i.e., an increased number of false negatives) with it.

The behavior of our classifiers in relation to changes in their thresholds can be well represented using ROC curves. These are plots of the true-positive versus the false-positive rate for a binary classifier as its discrimination threshold is varied.

To measure the performance of C_p on each of the two data sets, we used the AUC of the classifier output. However, these results can be biased in favor of the original BCI data, since the classifiers were derived from training data from D_0. Indeed, the addition of artifacts to create $D_{j,k}$ affects the features extracted and, consequently, can obscure the useful information that was originally available to the classifier when it derived a rule for predicting the label of a trial epoch. Moreover, the best set of electrodes found for the original BCI data set could be different from those found for the portable BCI data set (e.g., if the optimal subset of electrodes is calculated *after* adding the noise, the feature selection step will try to avoid selecting electrodes that are heavily affected by it).

Therefore, we augmented our analysis by training a number of classifiers $C_{p,j,k}$ with the training set (generated as described in Section 9.3.4) of the simulated portable BCI data sets, where the subscripts have the same meaning as above. For each of these, an optimal subset of three to nine electrodes was derived before training the classifiers.

Finally, we used each $C_{p,j,k}$ to predict the classes of the trials in the associated test set and the AUC to measure the performance. The results of this analysis are reported in Section 9.4.3.

9.4 RESULTS

9.4.1 BASELINE

In order to measure the extent to which a certain type of artifact affects the BCI, we first need to assess its performance under ideal laboratory conditions.

Table 9.1 shows the AUCs obtained using, for each classifier C_p, the training and test sets of participant $p \in [1,8]$ extracted from the BCI mouse data set D_0.

We can observe that, with the approach described above, we achieved lower AUC values than those reported in Ref. [14]. This is reasonable since we used a lower number of features and a less complex classifier (FDA instead of a linear support vector machine) than the original research, in order to simplify the BCI system. However, in this chapter, we focus on the differences in performance between original and artifact-contaminated data sets, and not on the absolute performance of the BCI system.

9.4.2 PERFORMANCE ON SIMULATED PORTABLE BCI DATA SETS

After measuring the performance of our BCI under ideal laboratory conditions, we simulated the case in which one would train the classifier with data collected in the laboratory, in nearly ideal conditions, and then test the portable BCI out of the

TABLE 9.1
AUC Values Obtained by Our BCI
for Each Participant for Target
versus Nontarget Classification

Participant	AUC
1	0.806
2	0.747
3	0.786
4	0.824
5	0.774
6	0.778
7	0.843
8	0.849
Mean	0.801
Median	0.796

Note: The last rows report the mean and median values across all participants.

laboratory. For this, we used the same classifiers C_p developed in Section 9.4.1 but calculated the AUC on the simulated portable BCI data sets ($D_{j,k}$).

In this way, we can study the effects that each type of noise has on the performance of the BCI. The results of this analysis are collected in Table 9.2, where each value represents the median AUC across the individual classifiers C_p when testing on data from participant $p \in [1,8]$ from the data set $D_{j,k}$. The numbers in brackets

TABLE 9.2
Performance of the BCI When Trained with Trials from D_0 and Tested on Epochs from $D_{j,k}$

Noise Type	$k = 1$	$k = 2$	$k = 3$	Rank
Turn neck to the left	0.775 (0.117)	**0.716** (0.003)	**0.706** (0.000)	22
Turn neck to the right	0.763 (0.065)	*0.761* (0.025)	*0.741* (0.010)	20
Move head up	*0.747* (0.019)	**0.736** (0.002)	**0.686** (0.000)	25
Move head down	0.769 (0.065)	**0.715** (0.001)	**0.692** (0.001)	24
Move your tongue (mouth closed)	0.780 (0.323)	*0.759* (0.025)	*0.753* (0.032)	12
Move eyes to the left	0.766 (0.065)	0.768 (0.032)	0.769 (0.025)	8
Blink once	0.786 (0.439)	0.745 (0.080)	*0.768* (0.014)	10
Blink repeatedly	0.793 (0.287)	*0.768* (0.014)	*0.763* (0.032)	2
Move eyes to the right	0.774 (0.139)	0.762 (0.080)	*0.752* (0.025)	13
Look up	0.785 (0.334)	0.758 (0.052)	**0.739** (0.002)	17
Look down	0.778 (0.253)	0.761 (0.052)	*0.731* (0.019)	18
Turn the wheel left	0.773 (0.171)	0.776 (0.097)	**0.752** (0.007)	9
Turn the wheel right	0.784 (0.191)	*0.761* (0.010)	*0.743* (0.019)	15
Change gear with left hand	0.785 (0.171)	0.775 (0.080)	*0.756* (0.041)	5
Change gear with right hand	0.793 (0.221)	0.766 (0.117)	**0.747** (0.007)	7
Do nothing	0.789 (0.323)	*0.763* (0.032)	0.773 (0.080)	1
Yawn	0.786 (0.253)	*0.750* (0.014)	**0.747** (0.005)	16
Swallow	0.772 (0.080)	*0.761* (0.035)	**0.758** (0.007)	11
Count in loud voice	0.784 (0.360)	0.774 (0.065)	*0.764* (0.032)	3
Cross left foot over right	0.780 (0.262)	**0.749** (0.005)	*0.758* (0.032)	14
Cross right foot over left	0.780 (0.191)	*0.756* (0.019)	**0.730** (0.003)	19
Squint	0.787 (0.287)	0.765 (0.080)	*0.755* (0.032)	6
Bend neck to the right	0.775 (0.052)	**0.715** (0.000)	**0.696** (0.000)	23
Bend neck to the left	0.765 (0.065)	**0.735** (0.003)	**0.726** (0.000)	21
Electrical noise	0.778 (0.145)	*0.772* (0.032)	*0.769* (0.032)	4
Mean	0.778	0.755	0.743	

Note: Each number represents the median AUC across all participants for a specific type of artifact (from top to bottom, $j = 1, ..., 25$) and the given number of occurrences ($k = 1, ..., 3$). The last row represents the average AUC across data sets $D_{j \in [1,25], k}$. Numbers in parentheses represent the p values of a one-sided Wilcoxon test comparing BCI performance of classifiers C_p when tested on D_0 versus the same classifiers tested on $D_{j,k}$. Numbers in italics represent statistical significance at the 5% confidence level. Numbers in boldface represent statistical significance at the 1% confidence level. The last column represents the ranking table (low numbers represent higher AUCs) determined by the mean value across columns $k = 1, 2, 3$.

represent the p values of a one-sided Wilcoxon test comparing BCI performance of classifiers C_p when tested on D_0 versus the same classifiers tested on $D_{j,k}$.

Figure 9.5 shows examples of real trial epochs (*after* preprocessing) extracted from D_0 (blue line) and corresponding artifact-contaminated epochs from $D_{j,k}$ (black line). In this figure, we represent the effects of different types of artifacts that are approximately at the 90th percentile ("Blink once", at the top of the figure), 50th percentile ("Look down", in the middle), and 10th percentile ("Bend neck to the right", at the bottom) of the sample for $k = 3$. Hence, these examples represent the cases where the BCI is least, average, and most affected by noise, respectively.

In order to explain the results obtained, we created a ranking system where the first positions are those for which the average across columns $k = 1, 2, 3$ is highest. The first position is then taken by the action "Do nothing" (as expected). In this case, the type of noise that is being added to the BCI data set is an EEG signal with no other source of contamination. Thus, it is reasonable that the performance of the simulated portable BCI is not largely affected by the addition of this artifact, even though different electrode impedances will result in differences in signal amplitude, since the data have not been normalized.

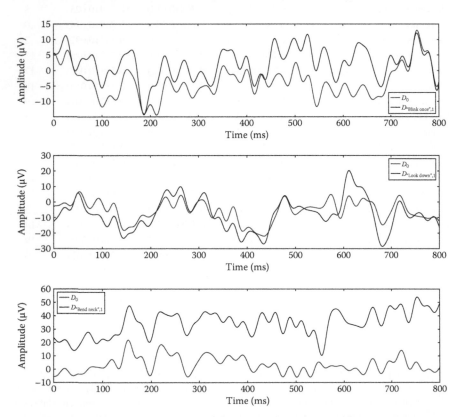

FIGURE 9.5 Examples of epochs from D_0 and $D_{j,k}$. Top: target trial with an eye blink at electrode site Cz; middle: target trial with an eye movement ("Look down") at Cz; bottom: target trial with a neck movement ("Bend neck to the right") at Oz.

Other artifacts that dominate the ranking are "Electrical noise" and "Squint". Again, the contribution of these should not largely affect BCI performance, given that we are filtering the data above 40 Hz. Mains noise appears as a frequency component of 60 Hz, so our filter removes most of it. Similarly, EMG artifacts from muscle contractions, such as those produced by squinting, can be eliminated by low-pass filtering the signals above 30 Hz.

At the lower end of the ranking, we mostly find artifacts related to neck movements (e.g., those that involve bending or turning the neck). This is not an unexpected result for one main reason: during the data collection of artifacts, the volunteer had a head rest on the chair, so neck movements most likely also involved changes in electrode position, with associated sustained voltage shifts. This is clearly represented in Figure 9.5 (bottom), even if the artifact epoch was preprocessed before adding it to the direction epoch. However, head rests are also available on a car; hence, the results obtained with our simulated portable BCI data sets are likely to be very similar to real data collected from a driver.

The last row of Table 9.2 represents the average value of each column, so that the effects of increasing the interference of artifacts on the BCI data can be observed in this summary. As expected, performance decreases monotonically with increasing values of k (i.e., when more noise is present, the number of false positives and false negatives increases, and as a result, we obtain lower AUC values). Even though this decrease is not statistically significant for $k = 1$ (only one type of artifact reaches statistical significance in this case), the effects of adding noise have a substantial impact on performance, as shown in the last column of the table, where for 24 out of the 25 types of artifacts (the only exception is the "Do nothing" activity, as expected), the performance of the BCI is significantly worse than in the ideal conditions of the BCI mouse experiment.

This raises the question of whether it is worth including some noisy trials when training the classifiers, once the types of artifacts that will later be encountered during the use of the portable BCI are known.

9.4.3 Effects of Training with Simulated Portable BCI Data Sets

If one knows in advance the type of artifacts that are more common in a mobile application of a BCI, he or she could wonder whether it is better to include some noise of that type during training the classifier, in order to increase the robustness of the system, or if performance is better when the training set is composed only of data collected under ideal conditions.

In order to check this, we performed an extra experiment in which we trained and tested classifiers using data from the simulated portable BCI data sets only. In particular, for each tuple (j,k), where j is the type of artifact and k is the number of occurrences of that type of artifact in each direction epoch (and, thus, gives an idea of the percentage of contamination in the data set), we trained classifier $C_{p,j,k}$ with the training set of participant p from data set $D_{j,k}$ and then calculated the individual AUC with the test set of that participant from the same data set. Table 9.3 reports the median AUC across all participants obtained in this way. The numbers in brackets represent the result of a one-sided Wilcoxon test comparing the

TABLE 9.3
Performance of the BCI When Trained and Tested on the Simulated Mobile BCI Data Sets

Noise Type	$k = 1$	$k = 2$	$k = 3$	Rank
Turn neck to the left	**0.742** (0.002)	**0.741** (0.007)	**0.708** (0.001)	21
Turn neck to the right	*0.747* (0.032)	*0.750* (0.041)	**0.710** (0.001)	20
Move head up	**0.716** (0.002)	**0.686** (0.000)	**0.668** (0.000)	25
Move head down	0.749 (0.052)	**0.735** (0.007)	**0.700** (0.001)	23
Move your tongue (mouth closed)	0.777 (0.164)	*0.769* (0.041)	**0.749** (0.003)	3
Move eyes to the left	0.777 (0.139)	*0.747* (0.032)	*0.760* (0.032)	6
Blink once	0.766 (0.065)	*0.744* (0.010)	**0.734** (0.002)	17
Blink repeatedly	0.769 (0.080)	*0.750* (0.010)	**0.738** (0.005)	14
Move eyes to the right	0.769 (0.097)	*0.753* (0.014)	0.756 (0.020)	7
Look up	0.788 (0.287)	**0.729** (0.001)	0.725 (0.010)	18
Look down	0.766 (0.065)	0.765 (0.052)	**0.747** (0.005)	9
Turn the wheel left	0.782 (0.171)	0.770 (0.139)	*0.766* (0.014)	1
Turn the wheel right	0.779 (0.145)	*0.738* (0.032)	**0.728** (0.001)	16
Change gear with left hand	0.775 (0.139)	**0.759** (0.007)	*0.755* (0.019)	4
Change gear with right hand	0.779 (0.164)	0.777 (0.117)	*0.754* (0.019)	2
Do nothing	0.780 (0.097)	*0.766* (0.041)	**0.739** (0.002)	5
Yawn	0.768 (0.221)	**0.733** (0.002)	*0.761* (0.041)	11
Swallow	0.766 (0.097)	**0.752** (0.005)	**0.742** (0.007)	12
Count in loud voice	*0.754* (0.032)	0.753 (0.052)	**0.734** (0.005)	19
Cross left foot over right	0.766 (0.080)	*0.751* (0.010)	**0.731** (0.002)	15
Cross right foot over left	0.775 (0.164)	*0.740* (0.010)	0.762 (0.014)	8
Squint	0.777 (0.117)	*0.744* (0.010)	*0.737* (0.010)	13
Bend neck to the right	0.761 (0.117)	**0.687** (0.001)	**0.682** (0.000)	24
Bend neck to the left	0.759 (0.052)	**0.713** (0.003)	**0.713** (0.001)	22
Electrical noise	*0.761* (0.014)	**0.759** (0.007)	**0.749** (0.007)	10
Mean	0.766	0.744	0.734	

Note: Each number represents the median AUC across all participants for a specific type of artifact (from top to bottom, $j = 1, ..., 25$) and the given number of occurrences ($k = 1, ..., 3$). The last row represents the average AUC across data sets $D_{j \in [1,25],k}$. Numbers in parentheses represent the p values of a one-sided Wilcoxon test comparing BCI performance of classifiers C_p when tested on D_0 versus classifiers $C_{p,j,k}$ tested on data sets $D_{j,k}$. Numbers in italics represent statistical significance at the 5% confidence level. Numbers in boldface represent statistical significance at the 1% confidence level. The last column represents the ranking table (low numbers represent higher AUCs) determined by the mean value across columns $k = 1, 2, 3$.

AUCs from classifiers $C_{p,j,k}$ (described above) with those of the baseline case (from Section 9.4.1).

As in the previous case (Section 9.4.2), we expected the AUCs to drop significantly with respect to the baseline case, especially for larger values of k. Indeed, this result was replicated, and the p values reported in Table 9.3 show that the differences for this case versus the original BCI data are more statistically significant.

If we now compare Table 9.3 with Table 9.2, we see that the ranking column shows no differences in the last positions of the ranking, which are still dominated by neck movements. In general, the order of the ranking list is maintained across all artifacts. However, there are some types of artifacts for which large variations are registered. In particular, the main changes in the ranking order are given by eye movements (e.g., "Move eyes to the right", "Look down"), which are now higher on the list than previously. Conversely, the main decreases on the ranking are given by eye blinks (both for one blink and several blinks). This suggests that this method is more robust to some types of artifacts than the one presented in Section 9.4.2. However, despite the changes in ranking order, the mean AUCs from which the rankings were calculated are still higher in Table 9.2 (i.e., when the classifiers are trained using only BCI data).

To answer the question of whether training with artifact-contaminated data is more suitable for portable BCIs than training under ideal circumstances, we checked for significant differences between the results of Table 9.2 versus Table 9.3 with a one-sided paired Wilcoxon test (results not reported). According to this, the only case in which the latter is significantly worse than the former consistently across all values of k is the "Blink once" activity (p values for $k = 1, 2, 3$ are 0.04, 0.039, and 0.02 respectively). Thus, despite the last row of both tables having such different mean values, it seems that (except when including blinks) training with noisy data will not affect the performance of the portable BCI during its use. Indeed, only in 4 out of the 25 types of artifacts was the difference between both methods significant for $k = 3$.

When taken together, all the evidence seems to point out that the first method of those presented (i.e., training with laboratory data) is either equal or significantly superior to the second, especially given the fact that eye blinks cannot be totally avoided during BCI use. However, when making this decision, one should also bear in mind that a lot of work has been done in terms of artifact correction methods for eye blinks.

9.5 CONCLUSIONS AND FUTURE WORK

BCIs have a great potential both in and outside of the laboratories. However, for this technology to work outside the ideal laboratory conditions, it is necessary to make it robust to artifacts and interferences.

Much of the noise that contaminates EEG data comes directly from the user in the form of EMG and EOG artifacts. Hence, for a portable BCI, it is important to characterize these and develop effective artifact correction methods that deal with them without affecting overall BCI performance and speed.

In this chapter, we ranked 25 types of artifacts on the basis of their effect on BCI performance decay. These included several forms of face muscular activity (e.g., squinting, blinking, and yawning) and neck, leg, and arm movements that are representative of actions that occur naturally while driving. Noise was artificially added to real BCI data from an Essex BCI Mouse experiment.

We analyzed the rankings in two different scenarios: one that simulates the action of training the BCI under ideal laboratory conditions and another that trains the machine using artifact-contaminated data. Both systems were then tested on a noisy data set.

We showed that performance decreases with respect to ideal laboratory conditions, but the drop is not too severe, which allows for the use of portable BCIs. Furthermore, with respect to whether there is an advantage in purposefully contaminating the BCI training set with artifacts that will later be found on the move, our results suggest that this is a suboptimal approach. By adding noise to the training data set, the classifier is not able to learn the ground truth properly and, if given the same number of samples as in "ideal" conditions, performance overall will be lower during online use.

One of the main limitations of our work is the fact that artifacts were artificially added to the BCI data, rather than collected directly during a BCI experiment. One of the reasons for using this method was that it allowed for more control over the data set. The collection of artifacts took approximately 2 h, which would have been a very long time for a pure BCI experiment. Also, by collecting data separately, we could create a variety of data sets, which would not have been possible otherwise (because of contamination of neural signals caused by the BCI paradigm and the extremely long time that it would have taken). Moreover, if artifacts and BCI data are collected together, participants of the experiment are given a dual task, which would also have negatively affected BCI performance for the baseline case.

Another limitation is the fact that artifacts were collected with the volunteer in a seated position. While these results might be generalized to the cases where the user is driving or screening for targets within bursts of images, we have not studied the effects of standing (and possibly walking) on BCI performance. The type of BCI that we tested in this work is a P300-based BCI. However, our approach can be generalized to other types of BCI (e.g., motor imagery or steady-state visual evoked potentials) and other EEG purposes like those mentioned in Section 9.1. In most types of existing BCIs, the user needs some degree of sustained attention to control the system. Hence, for instance, a person will not be able to use a P300-based BCI while walking. However, for portable EEG monitoring and EEG-based wearables, the effects of leg motion artifacts still need to be studied.

Despite the limitations discussed above, our results show that the effects of several types of noise on BCI performance can be greatly diminished with appropriate filtering. Moreover, the algorithms that we used for data preprocessing are fast enough to allow for online (i.e., real time) use of the BCI. They are also suitable to be implemented in a portable BCI system, which we will test in the future. Furthermore, despite their simplicity, we have shown that, for low levels of contamination, the decrease of performance of the BCI was not significant with respect to ideal laboratory conditions.

REFERENCES

1. Andrew B Schwartz. 2004. Cortical neural prosthetics. *Annual Review of Neuroscience*, 27:487–507.
2. Wilson Truccolo, Gerhard M Friehs, John P Donoghue, and Leigh R Hochberg. 2008. Primary motor cortex tuning to intended movement kinematics in humans with tetraplegia. *The Journal of Neuroscience: The Official Journal of the Society for Neuroscience*, 28(5):1163–1178, Jan.

3. Lawrence A Farwell and Emanuel Donchin. 1988. Talking off the top of your head: Toward a mental prosthesis utilizing event-related brain potentials. *Electroencephalography and Clinical Neurophysiology*, 70(6):510–523, Dec.

4. Jonathan R Wolpaw, Dennis J McFarland, Gregory W Neat, and Catherine A Forneris. 1991. An EEG-based brain–computer interface for cursor control. *Electroencephalography and Clinical Neurophysiology*, 78(3):252–259, Mar.

5. Gert Pfurtscheller, Doris Flotzinger, and Joachim Kalcher. 1993. Brain–computer interface: A new communication device for handicapped persons. *Journal of Microcomputer Applications*, 16(3):293–299.

6. Niels Birbaumer, Nimr Ghanayim, Thilo Hinterberger, Iver Iversen, Boris Kotchoubey, Andrea Kübler, Jouri Perelmouter, Edward Taub, and Herta Flor. 1999. A spelling device for the paralysed. *Nature*, 398(6725):297–298, Mar.

7. Nikolaus Weiskopf, Klaus Mathiak, Simon W Bock, Frank Scharnowski, Ralf Veit, Wolfgang Grodd, Rainer Goebel, and Niels Birbaumer. 2004. Principles of a brain–computer interface (BCI) based on real-time functional magnetic resonance imaging (fMRI). *IEEE Transactions on Biomedical Engineering*, 51(6):966–970, June.

8. Jonathan R Wolpaw and Dennis J McFarland. 2004. Control of a two-dimensional movement signal by a noninvasive brain–computer interface in humans. *Proceedings of the National Academy of Sciences of the U.S.A.*, 101(51):17849–17854, Dec.

9. Eric W Sellers and Emanuel Donchin. 2006. A P300-based brain–computer interface: Initial tests by ALS patients. *Clinical Neurophysiology*, 117(3):538–548, Mar.

10. Guido Dornhege, José del R Millán, Thilo Hinterberger, Dennis J McFarland, and Klaus-Robert Müller, editors. 2007. *Toward Brain-Computer Interfacing*. MIT Press, Cambridge, Massachusetts.

11. Roman Krepki, Gabriel Curio, Benjamin Blankertz, and Klaus-Robert Muller. 2007. Berlin brain–computer interface—The HCI communication channel for discovery. *International Journal of Human-Computer Studies*, 65(5):460–477.

12. Niels Birbaumer, Andrea Kübler, Nimr Ghanayim, Thilo Hinterberger, Jouri Perelmouter, Jochen Kaiser, Iver Iversen, Boris Kotchoubey, Nicola Neumann, and Herta Flor. 2000. The thought translation device (TTD) for completely paralyzed patients. *IEEE Transactions on Rehabilitation Engineering*, 8(2):191.

13. Jonathan R Wolpaw, Niels Birbaumer, William J Heetderks, Dennis J McFarland, P Hunter Peckham, Gerwin Schalk, Emanuel Donchin, Louis A Quatrano, Charles J Robinson, and Theresa M Vaughan. 2000. Brain–computer interface technology: A review of the first international meeting. *IEEE Transactions on Rehabilitation Engineering*, 8(2):164–173, June.

14. Mathew Salvaris, Caterina Cinel, Luca Citi, and Riccardo Poli. 2012. Novel protocols for P300-based brain–computer interfaces. *IEEE Transactions on Neural Systems and Rehabilitation Engineering*, 20(1):8–17.

15. Luca Citi, Riccardo Poli, Caterina. Cinel, and Francisco Sepulveda. 2008. P300-based BCI mouse with genetically-optimized analogue control. *IEEE Transactions on Neural Systems and Rehabilitation Engineering*, 16(1):51–61, Feb.

16. José del Ramón Millán, Rüdiger Rupp, Gernot R Müller-Putz, Roderick Murray-Smith, Claudio Giugliemma, Michael Tangermann, Carmen Vidaurre, Febo Cincotti, Andrea Kübler, Robert Leeb, Christa Neuper, Klaus-Robert Müller and Donatella Mattia. 2010. Combining brain–computer interfaces and assistive technologies: State-of-the-art and challenges. *Frontiers in Neuroscience*, (4):161. doi:10.3389/fnins.2010.00161.

17. Chin-Teng Lin, Hung-Yi Hsieh, Sheng-Fu Liang, Yu-Chieh Chen, and Li-Wei Ko. 2007. Development of a wireless embedded brain-computer interface and its application on drowsiness detection and warning. In Don Harris, editor, *Engineering Psychology and Cognitive Ergonomics*, volume 4562 of *Lecture Notes in Computer Science*, pages 561–567. Springer Berlin Heidelberg.

18. Danny Plass-Oude Bos, Boris Reuderink, Bram Laar, Hayrettin Grkk, Christian Mhl, Mannes Poel, Anton Nijholt, and Dirk Heylen. 2010. Brain-computer interfacing and games. In Desney S Tan and Anton Nijholt, editors, *Brain–Computer Interfaces*, Human-Computer Interaction Series, pages 149–178, Springer London.

19. Maria V Ruiz Blondet, Adarsha Badarinath, Chetan Khanna, and Zhanpeng Jin. 2013. A wearable real-time BCI system based on mobile cloud computing. In *Neural Engineering (NER), 2013 6th International IEEE/EMBS Conference on*, pages 739–742, San Diego, CA, Nov.

20. Mohammad Modarreszadeh and Robert N Schmidt. 1997. Wireless, 32-channel, EEG and epilepsy monitoring system. In *Engineering in Medicine and Biology Society, 1997. Proceedings of the 19th Annual International Conference of the IEEE*, volume 3, pages 1157–1160, Chicago, Oct.

21. Lars Torsvall, Torbjurn Akerstedt, Katja Gillander, and Anders Knutsson. 1989. Sleep on the night shift: 24-hour EEG monitoring of spontaneous sleep/wake behavior. *Psychophysiology*, 26(3):352–358.

22. David J Kupfer, F. Gordon Foster, Patricia Coble, Richard J McPartland, and Richard F Ulrich. 1978. The application of EEG sleep for the differential diagnosis of affective disorders. *American Journal of Psychiatry*, 135(1):69–74.

23. Ana Matran-Fernandez, Riccardo Poli, and Caterina Cinel. 2013. Collaborative brain–computer interfaces for the automatic classification of images. In *Neural Engineering (NER), 2013 6th International IEEE/EMBS Conference on*, pages 1096–1099, San Diego (CA), Nov 6–8, IEEE.

24. Ana Matran-Fernandez and Riccardo Poli. 2014. Collaborative brain–computer interfaces for target localisation in rapid serial visual presentation. In *Computer Science and Electronic Engineering Conference (CEEC), 2014 6th*, pages 127–132, Colchester, September 25–26 2014, IEEE.

25. Riccardo Poli, Davide Valeriani, and Caterina Cinel. 2014. Collaborative brain–computer interface for aiding decision-making. *PLoS ONE*, 9(7):e102693.

26. Todd C. Handy, editor. 2004. *Event-Related Potentials. A Method Handbook*. MIT Press, Cambridge, Massachusetts.

27. Steven J. Luck. 2005. *An Introduction to the Event-Related Potential Technique*. MIT Press, Cambridge, Massachusetts.

28. Riccardo Poli, Luca Citi, Francisco Sepulveda, and Caterina Cinel. 2009. Analogue evolutionary brain computer interfaces. *IEEE Computational Intelligence Magazine*, 4(4):27–31, Nov.

29. Mathew Salvaris, Caterina Cinel, Riccardo Poli, Luca Citi, and Francisco Sepulveda. 2010. Exploring multiple protocols for a brain–computer interface mouse. In *Proceedings of 32nd IEEE EMBS Conference*, pages 4189–4192, Buenos Aires, Sept.

30. Mehrdad Fatourechi, Ali Bashashati, Rabab K Ward, and Gary E Birch. 2007. EMG and EOG artifacts in brain computer interface systems: A survey. *Clinical Neurophysiology*, 118(3):480–494.

31. Rolf Verleger, Theo Gasser, and Joachim Möcks. 1982. Correction of EOG artifacts in event-related potentials of the EEG: Aspects of reliability and validity. *Psychophysiology*, 19(4):472–480, July.

32. Steven J Luck. 2014. *An Introduction to the Event-Related Potential Technique*. MIT Press, Cambridge, Massachusetts.

33. Aapo Hyvärinen, Juha Karhunen, and Erkki Oja. 2001. *Independent Component Analysis*. New York, Wiley-IEEE.

34. Scott Makeig, Anthony J. Bell, Tzyy-Ping Jung, and Terrence J. Sejnowski. 1996. Independent component analysis of electroencephalographic data. In David S. Touretzky, Michael C. Mozer, and Michael E. Hasselmo, editors, *Advances in Neural Information Processing Systems*, volume 8, pages 145–151, MIT Press, Cambridge, Massachusetts.

35. Scott Makeig, Marissa Westerfield, Tzyy-Ping Jung, James Covington, Jeanne Townsend, Terrence J Sejnowski, and Eric Courchesne. 1999. Functionally independent components of the late positive event-related potential during visual spatial attention. *The Journal of Neuroscience*, 19(7):2665–2680, Apr.

36. Tzyy-Ping Jung, Scott Makeig, Marissa Westerfield, Jeanne Townsend, Eric Courchesne, and Terrence J Sejnowski. 2001. Analysis and visualization of single-trial event-related potentials. *Human Brain Mapping*, 14(3):166–185, Nov.

37. Scott Makeig, Marissa Westerfield, Tzyy-Ping Jung, S Enghoff, Jeanne Townsend, Eric Courchesne, and Terrence J Sejnowski. 2002. Dynamic brain sources of visual evoked responses. *Science (New York, N.Y.)*, 295(5555):690–694, Jan.

38. Hugh Nolan, Robert Whelan, and Richard B Reilly. 2010. FASTER: Fully automated statistical thresholding for EEG artifact rejection. *Journal of Neuroscience Methods*, 192(1):152–162, Sept.

39. Jessica D Bayliss and Dana H Ballard. 1999. Single trial P300 recognition in a virtual environment. In *Proceedings of the International Congress on Computational Intelligence: Methods & Applications*, June 22–25, Rochester, New York, USA. ICSC Academic Press.

40. Rodney J Croft and Robert J Barry. 2000. Removal of ocular artifact from the EEG: A review. *Clinical Neurophysiology*, 30:5–19.

41. Markus Junghöfer, Thomas Elbert, Don M Tucker, and Brigitte Rockstroh. 2000. Statistical control of artifacts in dense array EEG/MEG studies. *Psychophysiology*, 37:523–532.

42. Daniel Göhring, David Latotzky, Miao Wang, and Raúl Rojas. 2013. Semi-autonomous car control using brain computer interfaces. In Sukhan Lee, Hyungsuck Cho, Kwang-JoonYoon and Jangmyung Lee, editors, *Intelligent Autonomous Systems 12*, pages 393–408, Berlin, Springer.

43. Riccardo Poli, Caterina Cinel, Ana Matran-Fernandez, Francisco Sepulveda, and Adrian Stoica. 2013. Towards cooperative brain–computer interfaces for space navigation. In *Proceedings of the 2013 International Conference on Intelligent User Interfaces*, pages 149–160, Santa Monica, CA, ACM, March.

10 An Advanced Insulin Bolus Calculator for Type 1 Diabetes

Peter Pesl, Pau Herrero, Monika Reddy,
Maria Xenou, Nick Oliver, and Pantelis Georgiou

CONTENTS

10.1 INTRODUCTION

10.1.1 DIABETES MELLITUS

Diabetes mellitus is a chronic condition that occurs either when the pancreas can no longer produce sufficient insulin (type 1) or when the body cannot process the insulin it does produce (type 2). The emerging absolute or relative deficiency of insulin results in a loss of glucose homeostasis, which causes hyperglycemia (high blood glucose levels). Symptoms of high glucose levels include polydipsia, polyuria, polyphagia, blurred vision, tiredness, and loss of weight. The World Health Organization has reported that an estimate of 171 million people suffered from diabetes in 2000, expecting this number to be doubled in 2030 [1]. Type 1 diabetes accounts for approximately 15% of all cases.

Large intervention trials performed in the European Union [2] and the United States [3] showed how tight glycemic control avoiding hyperglycemia could prevent long-term complications such as microvascular (retinopathy, nephropathy, and neuropathy) and macrovascular disease (cardiovascular disease, stroke, and peripheral vascular disease). Such complications are reported to place a heavy burden on health services [4]. These trials also reported the associated risk of induced hypoglycemia (low blood glucose) that underscores the crucial need for exact and timely insulin dosage [5]. Severe or prolonged hypoglycemia is a major concern for people with diabetes and can result in seizures, cardiac arrhythmias, and the "dead-in-bed" syndrome.

10.1.2 CURRENT INSULIN THERAPY

At the moment, people with type 1 diabetes mellitus (T1DM) and people with type 2 diabetes mellitus (T2DM) who depend on exogenous insulin self-manage their diabetes by checking their capillary blood glucose levels using finger-prick blood glucose meters and then injecting themselves with insulin several times a day [4]. Recent developments in diabetes management include continuous glucose monitors (CGMs) and continuous subcutaneous insulin infusion (CSII) pumps, which aim to support people with diabetes to achieve better glucose control. CGM devices continuously sample (every 5 min) glucose concentration in the interstitial fluid from a disposable glucose sensor, which is placed just under the skin. According to the Juvenile Diabetes Research Foundation Continous Glucose Monitoring Study Group [6], CGM devices help people with T1DM to achieve greater average glucose control (A1C) without increasing hypoglycemia. CSII pumps are body-worn devices with the size of a pager that are able to deliver insulin to the subcutaneous tissue. Connected to the pump is an infusion set, consisting of a small plastic tube and a soft cannula that is inserted under the user's skin. CSII pumps can be used as an alternative for multiple daily injections (MDIs) with an insulin pen and provide continuous delivery of insulin for intensive insulin therapy. They replace the use of slow-acting insulin for basal needs by delivering a varied dose of fast-acting insulin continually throughout day and night, at a preset rate. Furthermore, it is possible to manually set an insulin bolus dose in order to cover a meal. It has been reported that continuous subcutaneous insulin infusion achieves better blood glucose control in type 1 diabetes compared to MDI therapy [7].

Despite these developments in diabetes technology, as well as improvements in diabetes education, people with T1DM still struggle to achieve optimal glycemic control and prevent long-term complications [8]. The main reason for this is the vast amount of parameters that affect the glucose regulatory system, which need to be considered when calculating how much insulin is needed for a meal. In practice, simple rules, mainly based upon empirical experience, are used to estimate the necessary amount of bolus insulin. Carbohydrate counting and insulin dose adjustment at mealtimes, aiming for a personalized postprandial glucose target, is a major component in supporting people with T1DM to self-manage effectively and is one of the principal learning objectives of structured education [9]. Calculation of insulin doses includes subjects to estimate the amount of ingested carbohydrate, knowledge of insulin/carbohydrate ratios and insulin sensitivity factors (which may vary throughout the day), and an ability to do divisions and subtractions in the head. Resources to support carbohydrate counting exist and are helpful, but calculating insulin doses remain challenging for many people with type 1 diabetes.

To overcome this, insulin bolus calculators [10] have been integrated into smart capillary blood glucose meters and insulin pumps, which have shown potential to improve glycemic control [11,12]. However, these basic algorithms rarely take into account the variable effect of exercise, stress, alcohol, intercurrent illness, endocrine changes, and other external factors. Furthermore, current bolus calculators rely heavily on initial tuning and do not adapt to the individual's response, thus limiting their effectiveness.

10.1.3 Artificial Pancreas

Long-term research goals in diabetes management include the development of technology such as a closed-loop insulin delivery system (i.e., artificial pancreas) [13], which is intended to totally, or partially, remove the subject with diabetes from the decision-making process, so that insulin dosing can be calculated and administered with no user intervention. However, despite significant progress [14], it will be many years until this technology becomes a reality and it may not be suitable for everybody (e.g., lack of acceptability) [15]. It is worth remarking that the artificial pancreas is unique in taking responsibility for glucose concentrations, and it is this transfer of trust that makes it different from a bolus calculator, which just provides a recommendation that can be accepted or not by the user. Thus, even with an artificial pancreas, people relying on basal-bolus insulin therapy will still need to be supported in making complex therapeutic decisions such as calculating the right insulin doses. Furthermore, current types of insulin do not act fast enough to cope with meals in real time; hence, many of the current artificial pancreas systems use premeal boluses computed by bolus calculators to achieve improved postprandial glucose control [16]. Therefore, there is a clear need to improve the performance of existing insulin decision support systems.

10.1.4 Decision Support for Insulin Dosing

The potential of using decision support systems for insulin dosing has been advocated since the early 1980s. Several computerized algorithms have been proposed [17–22] to recommend insulin dose adjustments based on mathematical models, empirical rules, or approaches used in control theory. One of the first implementations of these algorithms into a pocket computer were presented by Chanoch et al. [17] and Schiffrin et al. [18] where a set of rules was used for insulin dose adjustments. Peters et al. [23] clinically evaluated a wallet-sized device that utilizes a control algorithm and a statistical model to suggest adjustments to the insulin dose. The experimental group using the decision support system achieved similar, but not superior, metabolic performance compared to a control group, who received manual decision support by an educational team. Owens et al. [24] used Iterative Learning Control [25] to exploit the repetitive nature of the insulin therapy regimen of the diabetic patient. This algorithm, referred to as Run-to-Run, uses an update law that corrects the insulin-to-carbohydrate ratio of a bolus calculator for the next day. A pilot clinical study showed the efficacy of the Run-to-Run algorithm in T1DM subjects [26]. However, the Run-to-Run algorithm presents some limitations. First of all, it only acts when the pre-prandial glucose value is in a predefined range (e.g., 70–120 mg/dL). Second, the algorithm requires two blood glucose measurements (e.g., at 60 minutes and 90 minutes after the meal) to capture the post-prandial excursion. However, the post-prandial excursion profile is subject to the meal composition and these two measurements may not be representative anymore for certain meals. Finally, the assumption of the repetitive nature of the insulin therapy regimen of the person with diabetes is unrealistic due to perturbation such as physical activity, alcohol consumption, hormone cycles and psychological stress. Nowadays, insulin bolus calculators (see Section 10.2.1) are incorporated in commercially available insulin pumps [27] and within capillary blood glucose monitors [28], which allow basic fine-tuning of their parameters based on rules (e.g., reducing insulin dose by a preset percentage when exercising). Although the results of these algorithms or guidelines implemented within handheld devices are encouraging, they still require considerable user effort (i.e., people with T1DM and healthcare professionals) and lack the flexibility to react on the vast number of factors that influence the glycemic profile in daily life scenarios and thus achieve suboptimal performance.

10.2 CONCEPT AND APPROACH

In this work, we present an innovative insulin advisory system that aims to provide superior decision support compared to current insulin bolus calculators [10] without increasing the burden to people with T1DM and clinicians. The reason for such superiority is the ability to calculate personalized insulin doses through the use of continuous glucose data, combined with an advanced insulin-dosing algorithm for decision support.

The proposed decision support algorithm, referred to as Advanced Bolus Calculator for Diabetes (ABC4D), enhances a standard bolus calculator by means of case-based reasoning (CBR) [29], a consolidated artificial intelligence technique that is well suited for problems with partial or uncertain information. The algorithm has been validated in silico [30] using a Food and Drug Administration–approved T1DM simulator [31], and clinical trials to evaluate its therapeutic effectiveness and usability are under way.

For this purpose, ABC4D has been implemented on a platform that comprises two main components: a smartphone platform allowing manual input of glucose and variables affecting blood glucose levels (e.g., meal carbohydrate content, exercise) and providing real-time insulin bolus recommendations for people with T1DM (see Figure 10.1) and a clinical expert platform to supervise the automatic adaptations of the bolus calculator parameters [32].

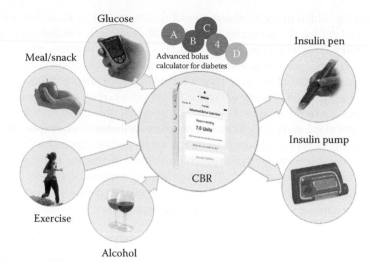

FIGURE 10.1 Graphical representation of the ABC4D (from left to right): Manual user inputs, smartphone holding the CBR algorithm and displaying personalized insulin advice, and insulin administration tools (insulin pen or insulin pump).

10.2.1 INSULIN BOLUS CALCULATORS

Insulin bolus calculators are simple decision support systems that consist of a relatively simple formula (see Equation 10.1) and use subject-specific metabolic parameters to calculate an insulin dose for a meal. The standard bolus calculator is described as

$$B = \frac{CHO}{ICR} + \frac{G - G^T}{ISF} - IOB, \tag{10.1}$$

where B is the recommended dose of insulin (IU) to be taken; CHO is the total amount of carbohydrate in the meal (g); ICR is the insulin-to-carbohydrate ratio (g/IU), which describes how many grams correspond to one unit of fast-acting insulin; G is the current capillary blood glucose level (mmol/L); G^T is the target blood glucose level (mmol/L); ISF is the insulin sensitivity factor (mg/L/IU), which is a personal relation describing how large a drop in blood glucose one unit of insulin gives rise to; and IOB is the insulin-on-board, which describes the amount of insulin still in the body from previous injections. Different formulas are being used by different manufacturers to estimate IOB (e.g., linear or curvilinear), with the linear estimation being the simplest one and described as

$$IOB = B_{k-1} \left(1 - \frac{t - T_B}{T_{IOB}} \right), \tag{10.2}$$

where B_{k-1} is the previously administered bolus, t is the time that IOB wants to be estimated, T_B is the time the bolus was administered, and T_{IOB} is a predefined interval during which the administered insulin is supposed to be active (e.g., 6 h). As mentioned earlier, the parameters ISF and ICR are usually not constant and may vary depending on external perturbations such as circadian rhythms, physical activity levels, hormone cycles, psychological stress, alcohol consumption, and recurrent illness. Although some of the most recent commercially available bolus calculators [28] allow consideration of some of these effects (e.g., exercise and hormone cycles), these features are rarely used because of the significant user effort that they require. Therefore, one of the main difficulties that people with diabetes, clinicians, and carers face when using bolus calculators is having to set its parameters and adjusting them on a regular basis according to changes in insulin requirements. Previous attempts have been carried out to automatically adjust the parameters of a bolus calculators [24] with varying success [26]. This is partly attributed to barriers such as requiring too many capillary blood glucose measurements or not contemplating realistic intrasubject variability.

In this work, we propose the utilization of CBR [33] do deal with the high variability in glycemic control and the use of CGM data to minimize the number of capillary blood glucose measurements.

10.2.2 CASE-BASED REASONING

CBR [33] is an artificial intelligence technique, which has been extensively applied in medicine [34]. CBR tries to solve newly encountered problems by applying the solutions learned from solving similar problems encountered in the past. This is equivalent to the way a human might solve a problem, that is, recalling a problem encountered in the past and, depending on whether the solution was successful or unsuccessful, either applying or avoiding that solution for the current situation.

In CBR, past situations are described by cases, which contain knowledge related to the various aspects of the situation. A case consists of three major parts: the problem description, the solution to the problem, and the outcome [33]. According to Aamodt and Plaza [29], general CBR consists of four steps, as summarized below and illustrated in Figure 10.2.

- **Retrieve** the most similar case(s) from a case base.
- **Reuse** the information and knowledge in that case(s) to solve the problem.
- **Revise** the proposed solution on the basis of the obtained outcome.
- **Retain** the parts of this experience likely to be useful for future problem solving.

The use of CBR for diabetes has been centred on the prognosis and risk of developing diabetes-related complications [35]. The first project to use CBR for recommending an insulin dose in diabetes management was the Telematic Management of Insulin-Dependent Diabetes Mellitus (T-IDDM) project [36], where a CBR engine was integrated with a rule-based reasoning engine and a probabilistic model of the effects of insulin on blood glucose levels. More recently, the 4 Diabetes Support System (4DSS) project [21] used CBR as the primary reasoning modality in a decision support tool for diabetics on insulin pump therapy and introduced other factors

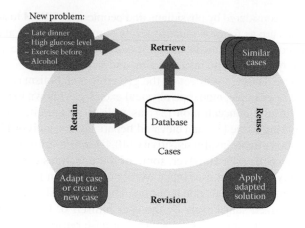

FIGURE 10.2 CBR cycle proposed by Aamodt and Plaza [29].

into the calculations, such as life events that can influence blood glucose levels. However, both systems were focused on providing decision support to the physicians using retrospective data and not real-time decision support for people with diabetes.

By shifting the focus on personalized insulin dosing decision support for people with T1DM, ABC4D utilizes CBR technology to introduce flexibility and adaptability to current insulin bolus calculators [30].

10.2.3 ABC4D ALGORITHM

This section describes the implementation of all four steps of the CBR cycle for the presented ABC4D system [32].

10.2.3.1 Case Definition

Cases are used to describe problem scenarios, where a solution is required. A case is defined by the triplet:

$$C := \{P, S, O\}, \tag{10.3}$$

where **P** is a set of parameters describing the problem (i.e., meal insulin dose calculation), **S** is the solution to the problem (i.e., insulin dose), and **O** is the outcome of applying the solution **S** to the problem described by **P** (i.e., postprandial excursion evaluation).

10.2.3.2 Case Parameters

Case parameters (**P**) are used to describe the current problem of calculating an insulin dose for a meal and compare the new scenario to previously used cases in a database (i.e., case base) for similarity. The number and importance of parameters play a key role in determining how similar a case is to the current problem. The amount of bolus insulin needed to bring glucose levels into a healthy range after a meal depends on various factors that affect the glucose and insulin dynamics. Some of the most significant features considered by physicians and people with T1DM to adjust insulin/carbohydrate ratios are as follows: glucose levels at meal time, type and duration of exercise, alcohol consumption, medication, time of day, work schedule, illness, stress, menstrual cycles, type of insulin, injection site, and sleep cycles.

For most of these factors, the effect on the glycemic metabolism is well described in the literature. For example, moderate physical exercise is known to result in a drop of basal plasma insulin concentration [37] as well as an amplification of glucose uptake by the working tissue [38] and elevated hepatic glycogenolysis [39]. Moderate alcohol consumption affects insulin sensitivity [40,41], while stress has been reported to induce hyperglycemia [42]. However, some of these features cannot be measured by sensors and rely on manual user input. While exercise can potentially be classified with accelerometers or heart rate monitors, environmental factors such as stress can only be determined subjectively by the user himself or herself. Therefore, the final decision on which parameters to include for case retrieval will be a trade-off between complexity, accuracy, and usability of the system.

10.2.3.3 Case Solution

For ABC4D, the solution **S** of a case is represented by the parameters ICR and ISF (see Equation 10.1) of the bolus calculator, which are adjusted to suggest the amount of insulin needed for a meal. It is important to note that some of the features listed above, such as the injection site or the type of insulin, do not affect the amount but the absorption of the administered insulin. Therefore, apart from the insulin dose, further potential solutions could include the time of insulin delivery (e.g., suggesting to take half of the bolus at meal time and the remaining bolus at a later stage), the injection site, or the shape of the delivered bolus.

10.2.3.4 Case Outcome

The outcome **O** of the solution is the glucose excursion after the insulin dose, which may be calculated from capillary blood glucose measurements or CGM data. The following metrics can be used to evaluate the performance of a case outcome: discrete postprandial blood glucose levels [24], glucose area under the curve [30], relative glucose increment, time to peak of glucose excursion, or the postprandial glucose minima. For most of these metrics, continuous data from a CGM need to be available.

ABC4D assesses the outcome using the minimum postprandial glucose value [43] obtained by a CGM as seen in Figure 10.3.

10.2.3.5 Retrieving Step

As stated in Section 10.2.2, the first step of the CBR cycle consists of retrieving the most similar case from the case base. Many different retrieval techniques for cases can be found in the literature [44]. The retrieving mechanism chosen in the current application is based on a weighted average distance function (i.e., k-nearest neighbor) defined as

$$D = \frac{K_{P_1} d_{P_1} + \cdots + K_{P_j} d_{P_j} + \cdots + K_{P_n} d_{P_n}}{K_{P_1} + \cdots + K_{P_j} + \cdots + K_{P_n}}, \tag{10.4}$$

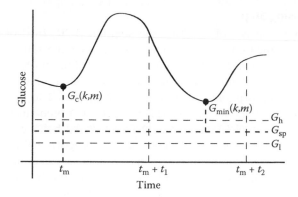

FIGURE 10.3   Graphical representation of the postprandial glucose excursion with the variables involved in the proposed metric. $G_c(k,m)$ is the capillary blood glucose value at the time of meal ingestion m. Horizontal dashed lines represent the glycemic target range ($[G_l,G_h]$) and the glucose set point (G_{sp}). Vertical thin dashed lines represent the time of the meal ingestion (t_m) and the postprandial time window ($[t_m + t_1, t_m + t_2]$) where minimum postprandial glucose value $G_{min}(k,m)$ is looked for.

where

$$d_{P_j} = \frac{abs(P_{j_k} - P_j)}{[P_j]},\qquad(10.5)$$

where P_j is a parameter from the current problem, P_{j_k} is the corresponding parameter of the retrieved case k from the case base, K_{P_j} is a weight associated to parameter P_j, which needs to be a priori specified by an expert and allows to assign the importance of a parameter on the retrieving procedure, and $[P_j]$ is the range of feasible values for P_j. Note that some of these ranges may be subject specific and need to be defined a priori based on retrospective data. Then, the case from the case library corresponding to the minimum distance D is the retrieved case.

10.2.3.6   Reusing Step

Reusing the retrieved solution to solve the current problem is done by applying the bolus calculator formula stated by Equation 10.1.

The retrieved solution can be further adjusted to better fit the current situation using a set of empirical rules. For example, a case has been retrieved that includes information about exercise after lunch. If no case exists in the case base describing the effect of exercise, then the closest case will be retrieved (i.e., lunch without exercise), which then needs to be adapted. Thus, the final recommended insulin dose is reduced by a predefined percentage as moderate exercise is known to increase insulin sensitivity. In a similar way, the solution of an exercise case can be adjusted, in case the user selects a higher intensity.

10.2.3.7 Revising Step

The revising step consists of updating the solution (ICR, ISF) of the retrieved case when the obtained glycemic outcome is not considered satisfactory. ABC4D utilizes the metric proposed in Ref. [43], which is based on run-to-run control [24] and uses CGM data to evaluate the minimum postprandial glucose measurement within a predefined postprandial time window (e.g., from 2 to 6 h postprandial). If the minimum postprandial glucose value is outside a predefined glycemic target range (e.g., 4–7 mmol/L), then an adjustment of the bolus calculator parameters ICR and ISF is required. Such a correction is proportional to the difference between the calculated minimum postprandial glucose value and the center of the glucose target range, which should correspond to the bolus calculator glucose set point (G_{sp}). The main advantage of the proposed metric is that it is more robust for individual and meal variability and does not require, compared to other methods [24], any initial tuning. The algorithm that implements such a metric is summarized as follows.

The minimum postprandial glucose value in a predefined time window is calculated as

$$G_{min}(k,m) = \min_{t \in [t_m + t_1, t_m + t_2]} G_{cgm}(t), \qquad (10.6)$$

where index k denotes the day (i.e., $k = 1, \ldots, n$), index m denotes the meal (i.e., breakfast, lunch, or dinner), G_{cgm} are the CGM measurements, t is the time variable, t_m is the time of the meal ingestion, and $t_m + t_1$ and $t_m + t_2$ are time instants defining the postprandial time window, where $t_1 < t_2$. It is important to remark that t_2 needs to be less or equal to the time of the next meal intake. In case additional insulin bolus is required to bring G_{min} into the predefined glycemic target range ($[G_l, G_h]$), then this can be calculated by

$$B_{add}(k,m) = \frac{(G_{min}(k,m) - G_{sp})}{ISF(k,m)}, \qquad (10.7)$$

where G_{sp} is the bolus calculator glucose set point calculated as $(G_h + G_l)/2$. Figure 10.3 shows a graphical representation of a postprandial excursion depicting the variables involved in the proposed metric. The correlation between ICR and ISF has been described in Ref. [45] and expressed by

$$ISF = \frac{1960\,ICR}{2.6W}, \qquad (10.8)$$

where W is the subject's weight in pounds. Then, by replacing Equation 10.8 into Equation 10.1 and isolating ICR, it is possible to calculate the adjusted ICR required to deliver the insulin dose ($B + IOB + B_{add}$) that brings G_{min} into the glycemic target range. That is,

$$ICR(k+1,m) = \frac{CHO(k,m) + \dfrac{G_c(k,m) - G_{sp}}{1960}}{\dfrac{2.6W}{B(k,m) + IOB(k,m) + B_{add}(k,m)}},$$ (10.9)

where index $k + 1$ denotes the updated value for the next day. The corresponding ISF($k + 1,m$) is then obtained by applying Equation 10.8. To make the adaptation more robust in front of noise and uncertainty, an average of the current and adjusted ICR is carried out.

10.2.3.8 Retaining Step

When no similar case to the problem being solved is found in the case base, a new case is created with a predefined solution, which is considered to be clinically safe but likely not optimal. This new case is then incorporated into the case base for further optimization and utilization. Another important aspect of the retention step is the maintenance of the case base. Maintenance strategies involve the detection and removal of redundant cases or multiple conflicting entries for the same situation.

10.2.3.9 Safety Constraints

To prevent insulin overdosing, two constraints were incorporated to avoid an excessive update of the solution (ICR). The first one limits the maximum allowed change of ICR for each revision by a preset percentage (e.g., 20%) while the second constraint defines the upper and lower limits that can be reached individually for each user.

10.3 IN SILICO EVALUATION

The commercial version of the UVa-Padova T1DM simulator [31] was used to evaluate the ABC4D algorithm for adjustment of the case solution (i.e., bolus calculator parameter ICR). In order to provide a more realistic scenario, we introduced variability in meal absorption, insulin absorption, insulin sensitivity, and uncertainty in glucose measurements [43]. Although three cases have been created for various meal times (i.e., breakfast, lunch, and dinner) to cope with intraday variations in insulin sensitivity, further cases could not be incorporated as the simulator does not incorporate perturbations such as physical exercise, illness, or stress. On the basis of evidence obtained from clinical trials evaluating the performance of existing bolus calculators [46], we compared the performance of the ABC4D algorithm to a standard bolus calculator without adaptation in a 3-month scenario. Table 10.1 shows simulation results comparing the glycemic outcomes employing three cases (i.e., breakfast, lunch, dinner) with and without case parameter adaptation on a cohort of 10 virtual adults (Table 10.1a) and on a cohort of 10 virtual adolescents (Table 10.1b). Results are expressed as mean ± SD.

In summary, the proposed adaptation method statistically improved ($p \leq 0.05$) all glycemic metrics evaluating hypoglycemia on both virtual cohorts: percentage time in hypoglycemia (i.e., BG < 3.9 mmol/L) (adults: 2.7 ± 4.0 vs. 0.4 ± 0.7, $p = 0.03$; adolescents: 7.1 ± 7.4 vs. 1.3 ± 2.4, $p = 0.02$) and low blood glucose index (LBGI) (adults: 1.1 ± 1.3 vs. 0.3 ± 0.2, $p = 0.002$; adolescents: 2.0 ± 2.2 vs. 0.7 ± 1.4, $p = 0.05$).

TABLE 10.1

Comparison of Three Bolus Calculator Cases (i.e., Breakfast, Lunch, and Dinner) with and without Parameter Adjustment on a Cohort of 10 Virtual Adults and 10 Virtual Adolescents

	Mean BG (mmol/L)	BG ∈ [3.9,10] mmol/L (% time)	BG < 3.9 mmol/L (% time)	BG > 10 mmol/L (% time)	LBGI
		(a) 10 Virtual Adults			
Without adjustment	7.2 ± 0.7	87.2 ± 13.9	2.7 ± 4.0	10.1 ± 10.5	1.1 ± 1.3
With adjustment	7.4 ± 0.6	90.1 ± 8.9	0.4 ± 0.7	9.4 ± 8.8	0.3 ± 0.2
p value	0.03[a]	0.5	0.03[a]	0.92	0.002[a]
		(b) 10 Virtual Adolescents			
Without adjustment	8.8 ± 0.9	61.7 ± 16.8	7.1 ± 7.4	31.2 ± 12.3	2.0 ± 2.2
With adjustment	8.8 ± 1.1	73.3 ± 18.3	1.3 ± 2.4	25.3 ± 16.6	0.7 ± 1.4
p value	0.92	0.16	0.02[a]	0.37	0.05[a]

[a] Statistically significant ($p < 0.05$).

A graphical example of a 3-month simulation run of the bolus calculator without and with adaptation mechanism on a virtual adolescent subject is shown in Figure 10.4. It is worth noting that, with the adaptation mechanism, the blue dots representing the minimum postprandial CGM values (G_{min}) are more concentrated in the glycemic target range ($[G_l, G_h]$) than without such mechanism. This is translated into a substantial improvement in glycemic control.

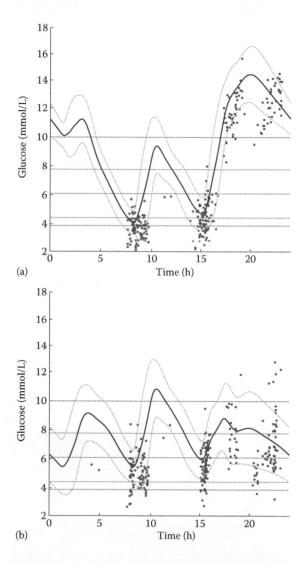

(a)

(b)

FIGURE 10.4 Graphical representation of a 3-month simulation run of the bolus calculator without and with adaptation mechanism on an adolescent subject. From down to top, horizontal lines represent hypoglycemic threshold, low target bound (G_l), mid target zone (G_{sp}), high target bound (G_h), and hyperglycemic threshold. The red solid line is the mean blood glucose and green dashed lines represent the standard deviation.

10.4   APPLICATION

ABC4D has been implemented into a smartphone platform running and a computer-based expert platform for clinical supervision (Figure 10.5). The smartphone platform provides real-time bolus advice to the user with T1DM, while the expert platform aims to guarantee safety by allowing a clinician to easily analyze and accept changes to the insulin therapy proposed by the CBR algorithm.

In order to ensure only safe case adaptations are performed, the CBR cycle is separated into two parts. The first part, comprising the retrieval and adaptation steps, is integrated into the smartphone platform, while the second part, containing the revision and the retention steps, is performed within the expert platform.

For practicality reasons, case revisions can be performed periodically by a clinical expert (e.g., weekly). Periodic revision also has the advantage of filtering out potential outliers if a case has been used more than once.

10.4.1   ABC4D SMARTPHONE PLATFORM

The ABC4D smartphone platform is implemented in an off-the-shelf smartphone (i.e., iPhone 4S, Apple Inc., California). Figure 10.5a (left) shows the main screen of the smartphone application used for requesting a new recommendation. It contains input elements to enter manual parameters (i.e., amount of carbohydrates, meal

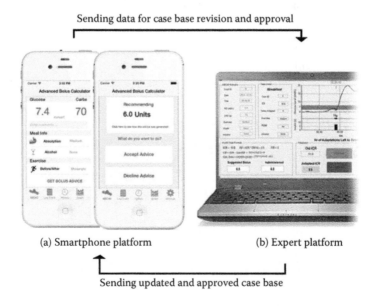

Sending data for case base revision and approval

(a) Smartphone platform   (b) Expert platform

Sending updated and approved case base

FIGURE 10.5 ABC4D smartphone platform (a) showing the main user interface on the left and displaying an insulin advice to the user on the right screen, respectively. All cases are periodically (e.g., once a week) sent to the expert platform (b) for revision. After all case adaptations have been approved by a clinical expert, the updated case base is sent back to the smartphone platform via encrypted e-mail.

absorption, current blood glucose level, alcohol consumption, and exercise) and a button for requesting insulin bolus advice.

The insulin recommendation is then presented to the user via a graphical user interface. Optionally, the user can see in detail how the advice has been generated by clicking on the insulin advice. Each recommendation needs to be accepted or declined manually, while the latter option requires the user to input the actual insulin dose that has been delivered. Additionally, the main menu enables the user to view logged data such as recent glucose levels, meal/exercise events, and previous insulin dose advices. If users have snacks in between main meals, they will have the option to introduce this information in the graphical user interface; however, no insulin recommendation will be given.

10.4.2 ABC4D EXPERT PLATFORM

The clinical expert platform runs on a desktop computer and is used to periodically perform case revisions by a clinician. First, all glucose information has to be extracted from the CGM and uploaded into the revision software. Then, all data essential for case revisions are sent from the smartphone to the expert platform. The ABC4D software on the smartphone enables users to send logbook and case data via encrypted e-mail to the expert performing the revision. Information from the logbook is needed because, in some scenarios, a case would not be revised. For instance, the user administered additional insulin or had a snack shortly after accepting an insulin advice. As these scenarios would affect the glucose profile after the meal, the outcome of the recommended insulin cannot be assessed. Figure 10.6 shows a

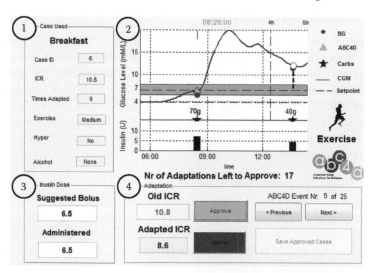

FIGURE 10.6 Example screenshot of the clinical revision software reviewing the performance of a retrieved case (1) that was used to describe a breakfast scenario with exercise (2). On the bottom of the screen, the clinical expert sees the suggested insulin advice (3) and the proposed adaptation to the insulin/carbohydrate ratio (ICR) of the case (4), which needs to be either approved or declined.

screenshot of the expert software, displaying the revision of proposed adaptation for the solution (i.e., ICR and ISF) of a presented case. After the revision of all cases has been completed, the software shows a summary of all adapted cases to the clinician. If one case has been used and revised multiple times, an average of all adaptations is calculated, which, in turn, needs to be manually approved. Finally, the case base on the smartphone platform needs to be updated. This can be performed either on the phone itself through an authorized settings menu or remotely via e-mail.

10.5 CLINICAL EVALUATION

Clinical safety and acceptability of ABC4D was evaluated as part of a clinical feasibility study ($n = 10$) over a period of 6 weeks. Inclusion criteria for participants were diagnosis of T1DM for >1 year, age >18 year, HbA1c <86 mmol/mol (<10%), and structured education completed. Ten adults with T1DM with a mean (SD) age of 42 (17) years, diabetes duration of 21 (15) years, and HbA1c = 64 (15) mmol/mol participated in the study.

All participants used the smartphone platform continuously for 6 weeks to receive insulin bolus advice and log additional diabetes events such as snacks, additional capillary blood glucose measurements, and correction boluses of insulin. After completion of the study, the participants were asked to complete a questionnaire assessing the usability of the system. A physician, a clinical research nurse, and an engineer used the expert platform at the end of each week throughout the 6-week study to approve case adaptations for each accepted bolus advice.

10.5.1 Usability and Acceptability

Human factors are key components to ensure adherence of users to information technologies for therapeutic purposes. For maximum performance, decision support systems for insulin dosing need to be as user-friendly as possible for both people with T1DM and clinicians. Results from the pilot study over 6 weeks evaluating the usability of ABC4D are encouraging, where 90% of all bolus recommendations were adhered to by the participants [32]. Furthermore, all participants found ABC4D to be user-friendly and nine out of ten participants stated to be happy to use the system for bolus recommendations.

10.5.1.1 Usage Frequency and Time

Special focus has been put on the usage of the smartphone platform over time, as it is expected that users will be more engaged at the beginning of the study and less so after several weeks. We found that participants used the logbook less at the end of the study phase to enter daily diabetes-related events (e.g., snacks, exercise, or stress) [32]. However, the number of insulin advice requests did not significantly change over the study period.

Although the time needed to request an insulin advice was significantly reduced in the last week compared to the start of the study, some participants still found the use of ABC4D more time consuming compared to their conventional way of calculating the insulin dose.

10.5.1.2 Safety and Trust

Safety and perceived safety are important aspects for the adoption of ABC4D. The proposed separation of the CBR cycle into a smartphone platform for insulin advice retrieval and a clinical expert platform for supervision ensures that all changes to the insulin therapy are clinically safe. After completion of the study, eight out of ten participants stated that they trusted the insulin advice that was generated by ABC4D.

10.6 DISCUSSION AND OUTLOOK

This work focuses on the development of an innovative decision support system for meal insulin dosing that provides enhanced adaptability and flexibility to current bolus calculators. The ultimate goal is that the intelligent bolus calculator will provide better glycemic control and reduce complications of diabetes. We presented an advanced bolus calculator (ABC4D) based on CBR that provides personalized insulin dosing for people with diabetes. Although the presented in silico results are encouraging, the simulation environment does not incorporate all of the uncertainty and perturbations that occur in the real world. Initial clinical studies have been completed assessing the feasibility and safety of the algorithm that has been implemented on a smartphone platform. Many people with T1DM struggle to achieve optimal glycemic control because of fear of hypoglycemia and its associated sequelae of reduced concentration levels, seizures, and cardiac arrhythmias. Integrating advanced diabetes technology, such as ABC4D, into day-to-day diabetes self-management has the potential to improve quality of life for people with diabetes.

The mobile implementation aims to further enhance the acceptability of the system, as many people with diabetes already carry smartphones with them, ensuring no additional devices are required. During development of the system, special focus has been put on the usability, but further improvements can still be achieved by reducing manual user input and therefore potentially increasing acceptability. To achieve this, ABC4D could potentially be integrated into a blood glucose meter or insulin pump to reduce the number of manual user inputs. Also, physical exercise could be measured using existing commercial devices such as heart rate monitors or accelerometers (e.g., Fitbit Inc., San Francisco, California).

It is important to note that the presented system can hold various algorithms for each of the CBR steps. The clinical performance of the whole system depends strongly on the implemented algorithms to retrieve, reuse, and adapt cases. Further studies are currently on-going to assess the clinical efficacy of ABC4D. In the mean time we show promising initial results of the presented system in a pilot study assessing usability and acceptability, which are key factors for the adoption of decision support systems for insulin dosing.

REFERENCES

1. S. Wild, G. Roglic, A. Green, R. Sicree, and H. King. 2004. Global prevalence of diabetes: Estimates for the year 2000 and projections for 2030. *Diabetes Care*, vol. 27, no. 5, pp. 1047–1053, May.

2. The UK Prospective Diabetes Study. 1998. Intensive blood-glucose control with sulphonylureas or insulin compared with conventional treatment and risk of complications in patients with type 2 diabetes (UKPDS 33). *The Lancet*, vol. 352, no. 9131, pp. 837–853.

3. The Diabetes Control and Complications Trial. 1993. The Effect of intensive treatment of diabetes on the development and progression of long-term complications in insulin-dependent diabetes mellitus. *New England Journal of Medicine*, vol. 329, pp. 977–986.

4. R. Holt, A. F. C. Cockram, and B. J. Goldstein. 2010. *Textbook of Diabetes: A Clinical Approach*, 4th Edition. Wiley-Blackwell.

5. DCCT. 1997. Hypoglycemia in the diabetes control and complications trial. *Diabetes*, vol. 46, no. 2, pp. 271–286.

6. ML Juvenile Diabetes Research Foundation Continuous Glucose Monitoring Study Group. "The effect of continuous glucose monitoring in well-controlled type 1 diabetes." Diabetes Care 32.8 (2009): 1378–1383.

7. J. Pickup, M. Mattock, and S. Kerry. 2002. Glycaemic control with continuous subcutaneous insulin infusion compared with intensive insulin injections in patients with type 1 diabetes: Meta-analysis of randomised controlled trials. *BMJ*, vol. 324, no. 7339, p. 705.

8. S. Adi. 2010. Type 1 diabetes mellitus in adolescents. *Adolescent Medicine: State of the Art Reviews*, vol. 21, no. 1, pp. 86–102.

9. Dafne Study Group et al. 2002. Training in flexible, intensive insulin management to enable dietary freedom in people with type 1 diabetes: Dose adjustment for normal eating (DAFNE) randomised controlled trial. *BMJ*, vol. 325, no. 7367, p. 746.

10. D. C. Klonoff. 2012. The current status of bolus calculator decision-support software. *Journal of Diabetes Science and Technology*, vol. 6, no. 5, pp. 990–994.

11. S. Schmidt, M. Meldgaard, N. Serifovski, C. Storm, T. M. Christensen, B. Gade-Rasmussen, and K. Nørgaard. 2012. Use of an automated bolus calculator in MDI-treated type 1 diabetes the boluscal study, a randomized controlled pilot study. *Diabetes Care*, vol. 35, no. 5, pp. 984–990.

12. C. Rees. 2014. Recommendations for insulin dose calculator risk management. *Journal of Diabetes Science and Technology*, vol. 8, no. 1, pp. 142–149.

13 A. Kowalski. 2009. Can we really close the loop and how soon? Accelerating the availability of an artificial pancreas: A roadmap to better diabetes outcomes. *Diabetes Technology & Therapeutics*, vol. 11, no. S1, pp. S–113.

14. S. J. Russell, F. El-Khatib, M. Sinha, K. Magyar, K. McKeon, L. Goergen, C. Balliro et al. 2014. Outpatient glycemic control with a bionic pancreas in type 1 diabetes. *New England Journal of Medicine*, vol. 371, no. 4, pp. 313–325.

15. K. Barnard, K. Hood, J. Weissberg-Benchell, C. Aldred, O. N., and L. L. 2014. Psychosocial assessment of artificial pancreas (AP): Commentary and review of existing measures and their applicability in AP research. *Diabetes Technology & Therapeutics*, vol. Epub ahead of print.

16. M. Reddy, P. Herrero, M. El Sharkawy, P. Pesl, N. Jugnee, H. Thomson, D. Pavitt et al. 2014. Feasibility study of a bio-inspired artificial pancreas in adults with type 1 diabetes. *Diabetes Technology & Therapeutics*, vol. 16, no. 9, pp. 550–557, Sept.

17. L. H. Chanoch, L. Jovanovic, and C. M. Peterson. 1985. The evaluation of a pocket computer as an aid to insulin dose determination by patients. *Diabetes Care*, vol. 8, no. 2, pp. 172–176.

18. A. Schiffrin, M. Mihic, B. S. Leibel, and A. M. Albisser. 1985. Computer-assisted insulin dosage adjustment. *Diabetes Care*, vol. 8, no. 6, pp. 545–552.

19. A. M. Albisser. 2003. Analysis: Toward algorithms in diabetes self-management. *Diabetes Technology & Therapeutics*, vol. 5, no. 3, pp. 371–373.

20, J. U. Poulsen, A. Avogaro, F. Chauchard, C. Cobelli, R. Johansson, L. Nita, M. Pogose et al. 2010. A diabetes management system empowering patients to reach optimised glucose control: From monitor to advisor. *Conference. Proceedings of the Annual International Conference of the IEEE Engineering in Medicine and Biology Society*, vol. 2010, pp. 5270–5271.

21. C. Marling, M. Wiley, T. Cooper, R. Bunescu, J. Shubrook, and F. Schwartz. 2011. The 4 diabetes support system: A case study in CBR research and development. In *Case-Based Reasoning Research and Development*, ser. *Lecture Notes in Computer Science*, A. Ram and N. Wiratunga, Eds. Springer Berlin Heidelberg, vol. 6880, pp. 137–150.

22. E. Salzsieder, L. Vogt, K.-D. Kohnert, P. Heinke, and P. Augstein. 2011. Model-based decision support in diabetes care. *Computer Methods and Programs in Biomedicine*, vol. 102, no. 2, pp. 206–218, May.

23. A. Peters, M. Rübsamen, U. Jacob, D. Look, and P. C. Scriba. 1991. Clinical evaluation of decision support system for insulin-dose adjustment in IDDM. *Diabetes Care*, vol. 14, no. 10, pp. 875–880.

24. C. L. Owens, H. Zisser, L. Jovanovic, B. Srinivasan, D. Bonvin, and F. J. Doyle, III. 2006. Run-to-run control of blood glucose concentrations for people with type 1 diabetes mellitus. *IEEE Transactions on Biomedical Engineering*, vol. 53, no. 6, pp. 996–1005.

25. F. Gao, Y. Wang, and F. J. Doyle III. 2009. Survey on iterative learning control, repetitive control, and run-to-run control. *Journal of Process Control*, vol. 19, no. 10, pp. 1589–1600.

26. C. C. Palerm, H. Zisser, W. C. Bevier, L. Jovanovič, and F. J. Doyle. 2007. Prandial insulin dosing using run-to-run control application of clinical data and medical expertise to define a suitable performance metric. *Diabetes Care*, vol. 30, no. 5, pp. 1131–1136.

27. H. Zisser, L. Robinson, W. Bevier, E. Dassau, C. Ellingsen, F. J. Doyle, and L. Jovanovic. 2008. Bolus calculator: A review of four "smart" insulin pumps. *Diabetes Technology & Therapeutics*, vol. 10, no. 6, pp. 441–444.

28. S. K. Garg, T. R. Bookout, K. K. McFann, W. C. Kelly, C. Beatson, S. L. Ellis, R. S. Gutin, and P. A. Gottlieb. 2008. Improved glycemic control in intensively treated adult subjects with type 1 diabetes using insulin guidance software. *Diabetes Technology & Therapeutics*, vol. 10, no. 5, pp. 369–375.

29. A. Aamodt and E. Plaza. 1994. Case-based reasoning: Foundational issues, methodological variations, and system approaches. *AI Communications*, vol. 7, no. 1, pp. 39–59.

30. P. Herrero, P. Pesl, M. Reddy, N. Oliver, P. Georgiou, and C. Toumazou. 2015. Advanced insulin bolus advisor based on run-to-run control and case-based reasoning. *IEEE Journal of Biomedical and Health Informatics*, vol. 19, no. 3, pp. 1087–1096.

31. B. P. Kovatchev, M. Breton, C. Dalla Man, and C. Cobelli. 2009. Biosimulation modeling for diabetes: In silico preclinical trials: A proof of concept in closed-loop control of type 1 diabetes. *Journal of Diabetes Science and Technology (online)*, vol. 3, no. 1, p. 44.

32. P. Pesl, P. Herrero, M. Reddy, M. Xenou, N. Oliver, D. Johnston, C. Toumazou, and P. Georgiou. 2015. An advanced bolus calculator for type 1 diabetes: System architecture and usability results. *IEEE Journal of Biomedical and Health Informatics*, vol. PP, no. 99, pp. 1–1 (online, ahead of print).

33. J. Kolodner. 1993. *Case-Based Reasoning*. Morgan Kaufmann Publishers, Inc.

34. R. Schmidt, A. Holt, I. Bichindaritz and P. Perner. 2006. Medical applications in case-based reasoning. *Cambridge University Press*, vol. 20, no. 3, pp. 289–292.

35. E. Armengol, A. Palaudries, and E. Plaza. 2001. Individual prognosis of diabetes long-term risks: A CBR approach. *Methods of Information in Medicine*, vol. 40, no. 1, pp. 46–51, Mar.

36. R. E. A. Bellazzi. 2002. A telemedicine support for diabetes management: The T-IDDM project. In *Computer Methods and Programs in Biomedicine*, vol. 69, no. 2, pp. 147–161.

37. P. Felig and J. Wahren. 1975. Fuel homeostasis in exercise. *New England Journal of Medicine*, vol. 293, no. 21, p. 1078.

38. D. H. Wasserman and A. D. Cherrington. 1991. Hepatic fuel metabolism during muscular work: Role and regulation. *American Journal of Physiology-Endocrinology and Metabolism*, vol. 260, no. 6, pp. E811–E824.

39. J. Wahren, P. Felig, G. Ahlborg, and L. Jorfeldt. 1971. Glucose metabolism during leg exercise in man. *Journal of Clinical Investigation*, vol. 50, no. 12, p. 2715.

40. A. Avogaro, R. M. Watanabe, L. Gottardo, S. de Kreutzenberg, A. Tiengo, and G. Pacini. 2002. Glucose tolerance during moderate alcohol intake: Insights on insulin action from glucose/lactate dynamics. *The Journal of Clinical Endocrinology & Metabolism*, vol. 87, no. 3, pp. 1233–1238.

41. S. Kiechl, J. Willeit, W. Poewe, G. Egger, F. Oberhollenzer, M. Muggeo, and E. Bonora. 1996. Insulin sensitivity and regular alcohol consumption: Large, prospective, cross sectional population study (Bruneck study). *BMJ*, vol. 313, no. 7064, pp. 1040–1044.

42. K. C. McCowen, A. Malhotra, and B. R. Bistrian. 2001. Stress-induced hyperglycemia. *Critical Care Clinics*, vol. 17, no. 1, pp. 107–124.

43. P. Herrero, P. Pesl, J. Bondia, M. Reddy, N. Oliver, P. Georgiou, and C. Toumazou. 2015. Method for automatic adjustment of an insulin bolus calculator: In silico robustness evaluation under intra-day variability. *Computer Methods and Programs in Biomedicine*, vol. 119, pp. 1–8, ISSN: 0169-2607.

44. P. Cunningham. 2009. Real-time hypoglycemia prediction suite using continuous glucose monitoring: A safety net for the artificial pancreas. *Knowledge and Data Engineering, IEEE Transactions on*, vol. 21, no. 11, pp. 1532–1543.

45. J. Walsh, R. Roberts, and T. Bailey. 2011. Guidelines for optimal bolus calculator settings in adults. *Journal of Diabetes Science and Technology*, vol. 5, no. 1, pp. 129–135.

46. G. Lepore, A. R. Dodesini, I. Nosari, C. Scaranna, A. Corsi, and R. Trevisan. 2012. Bolus calculator improves long-term metabolic control and reduces glucose variability in pump-treated patients with type 1 diabetes. *Nutrition, Metabolism and Cardiovascular Diseases*, vol. 22, no. 8, pp. e15–e16, Aug.

Index